Volume Doubling Measures and Heat Kernel Estimates on Self-Similar Sets

MEMOIRS
of the
American Mathematical Society

Number 932

Volume Doubling Measures and Heat Kernel Estimates on Self-Similar Sets

Jun Kigami

May 2009 • Volume 199 • Number 932 (third of 6 numbers) • ISSN 0065-9266

American Mathematical Society
Providence, Rhode Island

2000 *Mathematics Subject Classification.*
Primary 28A80, 60J35; Secondary 31C25, 60J45.

Library of Congress Cataloging-in-Publication Data

Kigami, Jun.
Volume doubling measures and heat kernel estimates on self-similar sets / Jun Kigami.
p. cm. — (Memoirs of the American Mathematical Society, ISSN 0065-9266 ; no. 932)
"Volume 199, Number 932 (third of 6 numbers)."
Includes bibliographical references and index.
ISBN 978-0-8218-4292-8 (alk. paper)
1. Measure theory. 2. Fractals. 3. Self-similar processes. I. Title.

QA312.K535 2009
515′.42—dc22 2008055058

Memoirs of the American Mathematical Society

This journal is devoted entirely to research in pure and applied mathematics.

Subscription information. The 2009 subscription begins with volume 197 and consists of six mailings, each containing one or more numbers. Subscription prices for 2009 are US$709 list, US$567 institutional member. A late charge of 10% of the subscription price will be imposed on orders received from nonmembers after January 1 of the subscription year. Subscribers outside the United States and India must pay a postage surcharge of US$65; subscribers in India must pay a postage surcharge of US$95. Expedited delivery to destinations in North America US$57; elsewhere US$160. Each number may be ordered separately; *please specify number* when ordering an individual number. For prices and titles of recently released numbers, see the New Publications sections of the *Notices of the American Mathematical Society.*

Back number information. For back issues see the *AMS Catalog of Publications.*

Subscriptions and orders should be addressed to the American Mathematical Society, P. O. Box 845904, Boston, MA 02284-5904, USA. *All orders must be accompanied by payment.* Other correspondence should be addressed to 201 Charles Street, Providence, RI 02904-2294, USA.

Memoirs of the American Mathematical Society (ISSN 0065-9266) is published bimonthly (each volume consisting usually of more than one number) by the American Mathematical Society at 201 Charles Street, Providence, RI 02904-2294, USA. Periodicals postage paid at Providence, RI. Postmaster: Send address changes to Memoirs, American Mathematical Society, 201 Charles Street, Providence, RI 02904-2294, USA.

This publication is indexed in *Science Citation Index*®, *SciSearch*®, *Research Alert*®,
CompuMath Citation Index®, *Current Contents*®/*Physical, Chemical & Earth Sciences.*
Printed in the United States of America.

∞ The paper used in this book is acid-free and falls within the guidelines
established to ensure permanence and durability.
Visit the AMS home page at `http://www.ams.org/`

10 9 8 7 6 5 4 3 2 1 14 13 12 11 10 09

To my family, Keita, Rie and Hiroko

Contents

Abstract

This paper studies the following three problems.
1. When does a measure on a self-similar set have the volume doubling property with respect to a given distance?
2. Is there any distance on a self-similar set under which the contraction mappings have the prescribed values of contractions ratios?
3. When does a heat kernel on a self-similar set associated with a self-similar Dirichlet form satisfy the Li-Yau type sub-Gaussian diagonal estimate?
Those three problems turns out to be closely related. We introduce a new class of self-similar set, called rationally ramified self-similar sets containing both the Sierpinski gasket and the (higher dimensional) Sierpinski carpet and give complete solutions of the above three problems for this class. In particular, the volume doubling property is shown to be equivalent to the upper Li-Yau type sub-Gaussian diagonal estimate of a heat kernel.

Received by the editor May 26, 2005; and in revised form September 18, 2006.
2000 *Mathematics Subject Classification.* Primary 28A80, 60J35; Secondary 31C25, 60J45.
Key words and phrases. self-similar set, volume doubling property, heat kernel.

Prologue

0.1. Introduction

This paper has originated from two naive questions about a self-similar set K. The first one is when a (self-similar) measure μ on K has the volume doubling property ((VD) for short) with respect to a given distance d. Let $B_r(x,d) = \{y|d(x,y) < r\}$ and let $V(x,r)$ be the volume of the ball $B_r(x,d)$, i.e. $V(x,r) = \mu(B_r(x,d))$. We say that μ has (VD) if and only if

$$V(x,2r) \le cV(x,r)$$

for any x and r, where c is independent of x and r. The simplest situation is when $V(x,r) = cr^n$ for any r and any x as we can observe in the case of the Lebesgue measures on the Euclidean spaces. Note that in such a case, $V(x,r)$ is homogeneous in space. The next best situation is to have (VD). Under it, we may allow inhomogeneity in space and, at the same time, still have good control of the volume by the distance. (VD) plays an important role in many area of analysis and geometry, for example, harmonic analysis, geometric measure theory, global analysis and so on.

The second question is when a heat kernel $p(t,x,y)$ on a self-similar set satisfies the following type of on-diagonal estimate

$$\frac{c_1}{V(x,t^{1/\beta})} \le p(t,x,x) \le \frac{c_2}{V(x,t^{1/\beta})}, \tag{0.1.1}$$

for $t \in (0,1]$. The estimate (0.1.1) immediately implies

$$\lim_{t\to 0} -\frac{\log p(t,x,x)}{\log t} = \lim_{r\to 0} \frac{1}{\beta}\frac{\log \mu(B_r(x,d))}{\log r}.$$

This relates the asymptotic behavior of the heat kernel to the multifractal analysis on the measure. (See Falconer [**12, 13**] about multifractal analysis.) Such a relation has been observed in [**20**] for post critically finite sets and in [**8**] for Sierpinski carpets.

Since Barlow and Perkins [**9**], there have been extensive results on heat kernels on self-similar sets. Mainly those works have focused on sub-Gaussian estimate

$$p(t,x,y) \approx c_1 t^{-d_s/2} \exp\left(-c_2\Big(\frac{d(x,y)^\beta}{t}\Big)^{1/(\beta-1)}\right), \tag{0.1.2}$$

where d_s is a positive constant called the spectral dimension, $d(\cdot,\cdot)$ is a distance and β is a constant with $\beta \ge 2$. This type of estimate has been first established for the "Brownian motion" on the Sierpinski gasket in [**9**]. Then it has been proven for nested fractals in [**33**], affine nested fractals in [**14**] and the Sierpinski carpets in [7]. Note that (0.1.2) gives a homogeneous on-diagonal estimate

$$c_1 t^{-d_s/2} \leq p(t,x,x) \leq c_2 t^{-d_s/2} \tag{0.1.3}$$

The homogeneous estimate (0.1.3) is known to require exact match between the measure μ and the form $(\mathcal{E},\mathcal{F})$. To be more precise, let K be the self-similar set associated with a family of contractions $\{F_i\}_{i=1,\ldots,N}$, i.e., $K = \cup_{i=1}^N F_i(K)$. We consider heat kernels associated with a self-similar Dirichlet form $(\mathcal{E},\mathcal{F})$, where $\mathcal{E}$ is the form and $\mathcal{F}$ is the domain of the form, under a self-similar measure μ with weight $\{\mu_i\}_{i\in S}$, where $S = \{1,\ldots,N\}$. $(\mathcal{E},\mathcal{F})$ is said to have self-similarity if

$$\mathcal{E}(u,v) = \sum_{i=1}^N \frac{1}{r_i}\mathcal{E}(u\circ F_i, v\circ F_i)$$

for any $u,v \in \mathcal{F}$, where $(r_1,\ldots,r_N)$ is a positive vector called resistance scaling ratio. Also a probability measure on K is called a self-similar measure on K with weight $\{\mu_i\}_{i\in S}$ if

$$\mu(A) = \sum_{i=1}^N \mu_i \mu(F_i^{-1}(A))$$

for any measurable set A. By the results in [**20, 28, 8**], the homogeneous on-diagonal estimate (0.1.3) holds if the ratio between $\log r_i$ and $\log \mu_i$ is independent of i. Otherwise, we may only expect inhomogeneous estimate (0.1.1) at the best.

The first and the second questions may look completely independent at a glance. They are, however, closely related. One of the main result in this paper is that the volume doubling property is equivalent to the upper inhomogeneous on-diagonal heat kernel estimate

$$p(t,x,x) \leq \frac{c}{V(x,t^{1/\beta})}. \tag{0.1.4}$$

for $t \in (0,1]$. Moreover, it turns out that the upper estimate (0.1.4) implies the upper and lower estimate (0.1.1). As a consequence, the first and the second questions are virtually the same. In fact, it has been known that (VD) combined with other properties is equivalent to the following Li-Yau type estimate of a heat kernel,

$$p(t,x,y) \approx \frac{c_1}{V(x,t^{1/\beta})}\exp\left(-c_2\left(\frac{d(x,y)^\beta}{t}\right)^{1/(\beta-1)}\right). \tag{0.1.5}$$

For example, in the case of Riemannian manifolds, Grigor'yan [**16**] and Saloff-Coste [**38**] have shown that (0.1.5) is equivalent to (VD) and the Poincaré inequality. See [**19, 17**] for other settings. In our case, the self-similarity of the space and the form allow (VD) itself to be equivalent to the heat kernel estimate (0.1.4).

At this point, a careful reader might notice that something is missing. Indeed, we have not mentioned what kind of distance we use in (0.1.1). In the course of our study, the natural distance for a heat kernel estimate like (0.1.1) should be a distance under which the system of contractions $\{F_i\}_{i\in S}$ has an asymptotic contraction ratio $\{(r_i\mu_i)^{\alpha/2}\}_{i\in S}$ for some α, i.e. $d(F_{w_1\ldots w_m}(x), F_{w_1\ldots w_m}(y))$ is asymptotically $(\gamma_{w_1}\cdots\gamma_{w_m})^\alpha d(x,y)$, where $\gamma_i = \sqrt{r_i\mu_i}$ and $F_{w_1\ldots w_m} = F_{w_1}\circ\ldots\circ F_{w_m}$ for $w_1,\ldots,w_m \in S$. Does such a distance really exist or not? Generalizing this, we have the third question. For a given ratio $\mathbf{a} = (a_i)_{i\in S}$, is there any distance under which $\{F_i\}_{i\in S}$ has the asymptotic contraction ratio $\mathbf{a}$? A similar problem has been studied in [**25**] for post critically finite self-similar sets. We will consider broader class of self-similar set with a different approach.

The key idea to study the third question is the notion of a scale, which essentially gives the size of $K_{w_1\ldots w_m} = F_{w_1\ldots w_m}(K)$. For a given ratio $\mathbf{a} = (a_1, \ldots, a_N)$, we think of $a_{w_1\ldots w_m} = a_{w_1}\cdots a_{w_m}$ as the size of $K_{w_1\ldots w_m}$. (Note that we do not suppose the existence of any distance at this point. If there were a distance which satisfies $d(F_i(x), F_i(y)) = a_i d(x,y)$, then the size of $K_{w_1\ldots w_m}$ had to be $a_{w_1\ldots w_m}$.) Starting from the scale (i.e. the size of $K_{w_1\ldots w_m}$), we will construct a system of fundamental neighborhoods $\{U_s(x)\}_{s\in(0,1]}$, which is the counterpart of balls with radius s and center x under a distance. See Section 1.3 for details. Now the problem is the existence of a distance whose balls match the virtual balls $\{U_s(x)\}$, or to be more exact, there is a distance d which satisfies

$$B_{c_1 s}(x,d) \subseteq U_s(x) \subseteq B_{c_2 s}(x,d) \tag{0.1.6}$$

for any s and any x or not, where c_1 and c_2 are independent of s and x. We say that a distance d is adapted to a scale if (0.1.6) holds.

As a whole, we will study three problems in this paper. Introduced in accordance with the appearance in this paper, they are

(P1) When does a (self-similar) measure have the volume doubling property with respect to a scale? The volume doubling property with respect to a scale means that

$$\mu(U_{2s}(x)) \leq c\mu(U_s(x))$$

for any $s \in (0, 1/2]$ and any $x \in K$.

(P2) Is there a good distance which is adapted to a given scale?

(P3) When does (0.1.1) hold for the heat kernel associated with a self-similar Dirichlet form and a (self-similar) measure?

(P1), (P2) and (P3) will be studied in Chapter 1, 2 and 3 respectively. Also those three questions are shown to be closely related in the course of discussion. In Chapter 1, we are going to introduce three properties, namely, an elliptic measure (EL), a locally finite scale (LF) and a gentle measure (GE). In short, (VD) turns out to be equivalent to the combination of (EL), (LF) and (GE). See Theorem 1.3.5. In the following sections, we will try to get simpler and effective description of (EL), (LF) and (GE) respectively for a restricted class of self-similar sets called rationally ramified self-similar sets. This class includes post critically finite self-similar sets, the cubes in $\mathbb{R}^n$ and the (higher dimensional) Sierpinski carpets. Also, for this class, we will give a complete answer to (P2) in Corollary 2.2.8, saying that, for a given ratio $\mathbf{a} = (a_1, \ldots, a_N)$, the scale associated with $\mathbf{a}$ satisfies (LF) if and only if there exists a distance which matches to the scale associated with the ratio $((a_i)^\alpha)_{i\in S}$ for some $\alpha > 0$. Based on those results, close relation between (P1), (P2) and (P3) will be revealed in Chapter 3. In particular, in Theorem 3.2.3, the following three conditions (a), (c) and (d) will be shown to be equivalent for rationally ramified self-similar sets:

(a) μ is (VD) with respect to the scale associated with the ratio $(\gamma_i)_{i\in S}$.

(c) $p(t,x,x) \leq \dfrac{c}{\mu(U_{\sqrt{t}}(x))}$ for $t \in (0,1]$.

(d) There exist $\alpha > 0$ and a distance d which is adapted to the scale associated with the ratio $((\gamma_i)^\alpha)_{i\in S}$ such that (0.1.4) holds, where $\beta = 2/\alpha$.

Moreover, if any of the above condition is satisfied, then we have full diagonal estimate (0.1.1) and the upper Li-Yau type estimate

$$p(t,x,y) \le \frac{c_1}{V(x,t^{1/\beta})} \exp\left(- c_2 \left(\frac{d(x,y)^\beta}{t} \right)^{1/(\beta-1)} \right). \tag{0.1.7}$$

for $t \in (0,1]$. Combining this results with the conclusion on (P2), we can easily determine self-similar measures for which (0.1.4) holds.

The organization of this paper is as follows. In Section 1.1, we introduce the notion of scales and establish several fundamental facts on this notion. In Section 1.2, we study self-similar structures and self-similar measures under the assumption that $K \neq \overline{V_0}$. This section gives bases of the discussions in the following sections. Section 1.3 is devoted to showing the equivalence between (VD) and the combination of (EL), (LF) and (GE) as we mentioned above. In Section 1.4, the properties (LF) and (GE) are closely examined. In particular, it is shown that (GE) is a equivalence relation among elliptic scales and (LF) is inherited by the equivalence relation (GE). The notion of rationally ramified self-similar set is introduced in Section 1.5. For this class of self-similar sets, we will find an effective and simple criteria for (LF) and (GE) in Section 1.6. We apply them to examples including post critically self-similar sets and the Sierpinski gasket in Section 1.7. The search of a distance which matches a scale starts at Section 2.1, where we define a pseudodistance associated with a scale. In Section 2.2, the notion of intersection type is introduced to give an answer to the existence problem of a distance adapted to a scale. Using the notion of qdistance, we will simplify the results in the previous two sections in Section 2.3. We will finally encounter with heat kernels in Section 3.1, which is completely devoted to setting up a reasonable framework of self-similar Dirichlet forms and the heat kernel associated with them. In Section 3.2, we establish a theorem to answer (P3), which will be the most important result in this paper. In Sections 3.3 and 3.4, we apply our main theorem to the post critically finite self-similar set and the Sierpinski carpets respectively. We need the entire Section 3.5 to complete the main theorem. In Appendixes, we mainly discuss relations between the properties of the heat kernel associated with a local regular Dirichlet from on a general measure-metric space.

0.2. the Unit square

Let us illustrate our main results by applying them to the unit square $[0,1]^2$, which is naturally self-similar. We denote the square by K and think of it as a subset of $\mathbb{C}$. Namely, $K = \{x + y\sqrt{-1} | x, y \in [0,1]\}$. The unit square can be regarded as a self-similar set in many ways. First, let $f_1(z) = z/2$, $f_2(z) = z/2+1/2$, $f_3(z) = z/2 + (1+\sqrt{-1})/2$ and $f_4(z) = z/2 + \sqrt{-1}/2$. Then $K = f_1(K) \cup f_2(K) \cup f_3(K) \cup f_4(K)$. According to the terminology in [**28**], K is the self-similar set with respect to $\{f_1, f_2, f_3, f_4\}$. K is not post critically finite but, so called, infinitely ramified self-similar set. Roughly speaking if any of $f_i(K) \cap f_j(K)$ is not a finite set, then K is called infinitely ramified self-similar set. In this case, $K_1 \cap K_2$ is a line, where $K_i = f_i(K)$.

Now let us explain the notion of "rationally ramified" self-similar sets by the unit square, which is the simplest (non trivial) rationally ramified self-similar set. There exists a natural map π from $\{1,2,3,4\}^{\mathbb{N}} \to K$ which is defined by $\pi(i_1 i_2 \ldots) = \cap_{m \ge 1} f_{i_1 \ldots i_m}(K)$. This map π determines the structure of K as a self-similar set. Note that the four line segments in the boundary of K is also self-similar sets. To

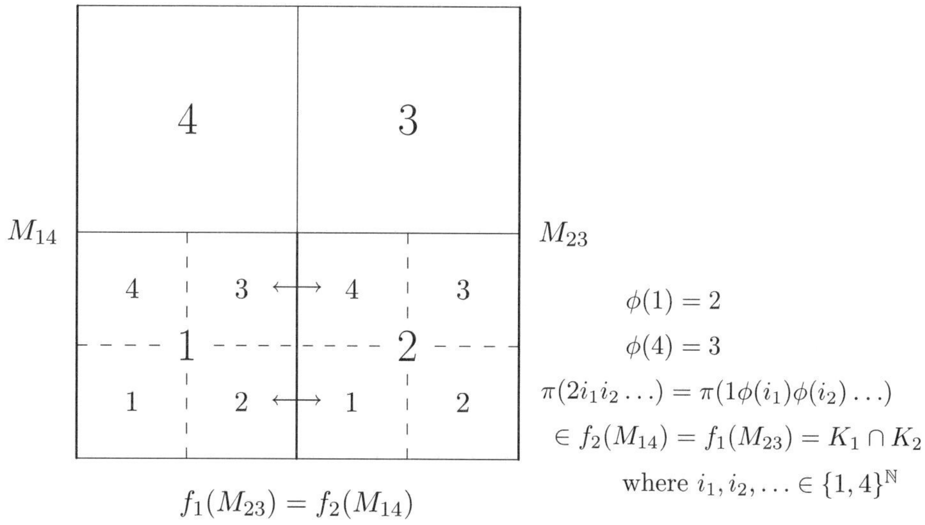

FIGURE 0.1. the square as a self-similar set

see this, set $M_{14} = \{\sqrt{-1}t | t \in [0,1]\}$ and $M_{23} = \{1 + \sqrt{-1}t | t \in [0,1]\}$ for example. Then $M_{14} = f_1(M_{14}) \cup f_4(M_{14})$ and $M_{23} = f_2(M_{23}) \cup f_3(M_{23})$ and hence M_{ij} is the self-similar set with respect to $\{f_i, f_j\}$. In other words, $M_{ij} = \pi(\{i,j\}^{\mathbb{N}})$. Those two self-similar sets M_{14} and M_{23} meet each other at $K_1 \cap K_2$ under the action of f_1 and f_2. More precisely, let $x \in K_1 \cap K_2$. Then there exists $i_1i_2\ldots \in \{1,4\}^{\mathbb{N}}$ such that $x = \pi(2i_1i_2\ldots) = \pi(1\phi(i_1)\phi(i_2)\ldots)$, where $\phi : \{1,4\} \to \{2,3\}$ is defined by $\phi(1) = 2, \phi(4) = 3$. See Figure 0.1.

Note that other intersections $K_i \cap K_j$ have similar descriptions. This is a typical example of rationally ramified self-similar set defined in 1.5, where an intersection of $f_i(K) \cap f_j(K)$ itself is a self-similar set and two different expressions (started from i and j respectively) by infinite sequences of symbols can be translated by a simple rewriting rules.

Next, applying the results in Chapter 1, we present the answer to the problem (P1) in this case. In particular, we can determine the class of self-similar measures which have the volume doubling property with respect to the Euclidean distance.

THEOREM 0.2.1. *A self-similar measure with weight* $(\mu_1, \mu_2, \mu_3, \mu_4)$ *has the volume doubling property with respect to the Euclidean distance if and only if* $\mu_1 = \mu_2 = \mu_3 = \mu_4 = 1/4$.

If $\mu_i = 1/4$ for all i, then μ is the restriction of the Lebesgue measure on K. So, the situation is very rigid and not quite interesting. In general, however, we can find richer structure of the volume doubling (self-similar) measures (even in the case of unit square). To see this, we are going to change the self-similar structure of the unit square.

From now on, K is regarded as a self-similar set with respect to nine contractions $\{F_i\}_{i=1,\ldots,9}$ in stead of four contractions $\{f_i\}_{i=1,\ldots,4}$ as above. Set $p_1 = 0, p_2 = 1/2, p_3 = 1, p_4 = 1+\sqrt{-1}/2, p_5 = 1+\sqrt{-1}, p_6 = 1/2+\sqrt{-1}, p_7 = \sqrt{-1}, p_8 = \sqrt{-1}/2$

a	b	a
c	d	c
a	b	a

$$a = a_1 = a_3 = a_5 = a_7$$
$$b = a_2 = a_6$$
$$c = a_4 = a_8$$
$$d = a_9$$

FIGURE 0.2. Weakly symmetric ratio

and $p_9 = 1/2 + \sqrt{-1}/2$. Define $F_i(z) = (z - p_i)/3 + p_i$ for $i = 1, \ldots, 9$. Then the square K is the self-similar set with respect to $\{F_i\}_{i \in S}$, where $S = \{1, \ldots, 9\}$, i.e. $K = \cup_{i \in S} F_i(K)$. In this case, we also have the natural map π from $S^{\mathbb{N}} = \{w_1 w_2 \ldots | w_i \in S\}$ to K defined by $\pi(w_1 w_2 \ldots) = \cap_{m \geq 0} F_{w_1 \ldots w_m}(K)$. Examining the intersection of $F_1(K)$ and $F_2(K)$, one may notice that $x = \pi(1 w_1 w_2 \ldots) = \pi(2 \varphi(w_1) \varphi(w_2) \ldots)$ for any $x \in F_1(K) \cap F_2(K)$, where $w_1 w_2 \ldots \in \{3, 4, 5\}^{\mathbb{N}}$ and $\varphi : \{3, 4, 5\} \to \{1, 8, 7\}$ is given by $\varphi(3) = 1, \varphi(4) = 8$ and $\varphi(5) = 7$. Also for any $y \in F_1(K) \cap F_8(K)$, we have $y = \pi(8 v_1 v_2 \ldots) = \pi(1 \psi(v_1) \psi(v_2) \ldots)$, where $v_1 v_2 \ldots \in \{1, 2, 3\}^{\mathbb{N}}$ and $\psi : \{1, 2, 3\} \to \{7, 6, 5\}$ is given by $\psi(1) = 7, \psi(2) = 6$ and $\psi(3) = 5$. This is again a typical example of a rationally ramified self-similar set.

Under this self-similar structure, self-similar volume doubling measures are much richer than before. The following condition will play an important role to solve all the three problems (P1), (P2) and (P3).

DEFINITION 0.2.2. A ratio $(a_i)_{i \in S} \in (0, 1)^S$ is called weakly symmetric if and only if $a_i = a_{\varphi(i)}$ for any $i \in \{3, 4, 5\}$ and $a_j = a_{\psi(j)}$ for any $j \in \{1, 2, 3\}$.

Note that a ratio $(a_i)_{i \in S}$ is weakly symmetric if and only if

$$a_1 = a_3 = a_5 = a_7, a_2 = a_6 \quad \text{and} \quad a_4 = a_8,$$

See Figure 0.2. First our results on (P1) in Chapter 1 yields the following characterization of the class of self-similar measures which are volume doubling with respect to the Euclidean distance.

THEOREM 0.2.3. *A self-similar measure with weight $(\mu_i)_{i \in S}$ has the volume doubling property with respect to the Euclidean distance if and only if $(\mu_i)_{i \in S}$ is weakly symmetric.*

As we have explained in the introduction, the main result of this paper is roughly the equivalence of the three properties: the volume doubling property of a measure, the existence of "asymptotically self-similar" distance and the upper and lower on-diagonal heat kernel estimate (0.1.1). In accordance with this spirit, being weakly symmetric gives an answer to (P2) as well. More precisely, the results in Chapter 2 gives the following theorem.

THEOREM 0.2.4. *Let $(a_i)_{i\in S} \in (0,1)^S$. $(a_i)_{i\in S}$ is weakly symmetric if and only if there exists a distance d which is adapted to the scale associated with a ratio $((a_i)^\alpha)_{i\in S}$ for some $\alpha > 0$.*

Naturally, weakly symmetric ratios appear again in our result on the problem (P3). To consider heat kernels, we regard K as a subset of $\mathbb{R}^2$ in the natural manner. Let ν be the restriction of the Lebesgue measure and let $\mathcal{F} = W^{1,2}(K)$. $W^{1,2}(K)$ is the Sobolev space defined by

$$W^{1,2}(K) = \{f | f \in L^2(K,\nu), \frac{\partial f}{\partial x}, \frac{\partial f}{\partial y} \in L^2(K,\nu)\},$$

where $\frac{\partial f}{\partial x}$ and $\frac{\partial f}{\partial y}$ are the partial derivatives in the sense of distribution. Note that ν is the self-similar measure with weight $(1/9, \ldots, 1/9)$. For any $f, g \in \mathcal{F}$, set

$$\mathcal{E}(f,g) = \int_K (\frac{\partial f}{\partial x}\frac{\partial g}{\partial x} + \frac{\partial f}{\partial y}\frac{\partial g}{\partial y}) dxdy.$$

Then $(\mathcal{E}, \mathcal{F})$ is a local regular Dirichlet form on $L^2(K,\nu)$ and the corresponding diffusion process is the Brownian motion which is reflected at the boundary of K. Moreover, the associated heat kernel satisfy the Gaussian type estimate

$$p(t,x,y) \asymp \frac{c_1}{t} \exp\Big(- c_2 \frac{|x-y|^2}{t}\Big)$$

for $t \in (0,1]$. This form $(\mathcal{E}, \mathcal{F})$ has the self-similarity with the resistance scaling ratio $(1, \ldots, 1)$, i.e.

$$\mathcal{E}(f,g) = \sum_{i\in S} \mathcal{E}(f \circ F_i, g \circ F_i).$$

for any $f, g \in \mathcal{F}$. Let μ be a self-similar measure with weight $(\mu_i)_{i\in S}$. Then by [**8**], making slight modifications, we may regard $(\mathcal{E}, \mathcal{F})$ as a local regular Dirichlet form on $L^2(K,\mu)$. At this time, the corresponding diffusion process is the time change of the Brownian motion. Let $p_\mu(t,x,y)$ be the associated heat kernel. (The heat kernel does exists and is jointly continuous in this case.) Then the results on (P3) implies the following.

THEOREM 0.2.5. *There exist $\alpha \in (0,1]$ and a distance d such that d is adapted to the scale associated with the ratio $((\mu_i)^{\alpha/2})_{i\in S}$ and the upper Li-Yau estimate (0.1.7) for $p_\mu(t,x,y)$ holds with $\beta = 2/\alpha$, if and only if $(\sqrt{\mu_i})_{i\in S}$ is weakly symmetric. Moreover, either of the above conditions suffices the upper and the lower on-diagonal estimate (0.1.1) for $p_\mu(t,x,y)$.*

CHAPTER 1

Scales and Volume Doubling Property of Measures

1.1. Scale

In this section, we introduce a notion of scales. A scale gives a fundamental system of neighborhoods of the shift space, which is the collection of infinite sequences of finite symbols. Later in Section 1.3, we will define a family of "balls" of a self-similar set through a scale.

NOTATION. For a set V, we define $\ell(V) = \{f | f : V \to \mathbb{R}\}$. If V is a finite set, $\ell(V)$ is considered to be equipped with the standard inner product $(\cdot, \cdot)_V$ defined by $(u, v)_V = \sum_{p \in V} u(p)v(p)$ for any $u, v \in \ell(V)$. Also $|u|_V = \sqrt{(u, u)_V}$ for any $u \in \ell(V)$.

Now we define basic notions on the word spaces and the shift space. Let S be a finite set.

DEFINITION 1.1.1. (1) For $m \geq 0$, the word space of length m, $W_m(S)$, is defined by

$$W_m(S) = S^m = \{w | w = w_1 \dots w_m, w_i \in S \,\text{for any}\ i = 1, \dots, m\}.$$

In particular $W_0(S) = \{\emptyset\}$, where $\emptyset$ is called the empty word. Also $W_*(S) = \cup_{m \geq 0} W_m(S)$ and $W_\#(S) = \cup_{m \geq 1} W_m(S)$. For $w \in W_m(S)$, we define $|w| = m$ and call it the length of the word w.
(2) For $w, v \in W_*(S)$, we define $wv \in W_*(S)$ by $wv = w_1 \dots w_m v_1 \dots v_n$, where $w = w_1 \dots w_m$ and $v = v_1 \dots v_n$. Also for $w^1, w^2 \in W_*(S)$, we write $w^1 \leq w^2$ if and only if $w^1 = w^2 v$ for some $v \in W_*(S)$.
(3) The (one sided) shift space $\Sigma(S)$ is defined by

$$\Sigma(S) = S^{\mathbb{N}} = \{\omega | \omega = \omega_1 \omega_2 \dots, \omega_i \in S \,\text{for any}\ i \geq 1\}.$$

The shift map $\sigma : \Sigma(S) \to \Sigma(S)$ is defined by $\sigma(\omega_1 \omega_2 \dots) = \omega_2 \omega_3 \dots$. For each $i \in S$, we define $\sigma_i : \Sigma(S) \to \Sigma(S)$ by $\sigma_i(\omega) = i\omega_1\omega_2 \dots$, where $\omega = \omega_1 \omega_2 \dots$. For $w = w_1 \dots w_m \in W_*(S)$, $\sigma_w = \sigma_{w_1} \circ \dots \circ \sigma_{w_m}$ and $\Sigma_w(S) = \sigma_w(\Sigma(S))$.
(4) The extended shift map $\sigma : W_*(S) \to W_*(S)$ is defined by $\sigma(\emptyset) = \emptyset$ and $\sigma(w_1 \dots w_m) = w_2 \dots w_m$ for any $w \in W_\#$. Also we extend $\sigma_i : W_*(S) \to W_*(S)$ by $\sigma_i(w_1 \dots w_m) = iw_1 \dots w_m$.

Note that $\leq$ is a partial order of $W_*(S)$. We write $w^1 < w^2$ if and only if $w^1 \leq w^2$ and $w^1 \neq w^2$. If no confusion can occur, we omit S in the above notations. For example, we write W_m in stead of $W_m(S)$.

The shift space Σ has a product topology as an infinite product of a finite set S. Under this topology, Σ is compact and metrizable. See [**28**] for details.

DEFINITION 1.1.2. (1) Let $\Lambda \subset W_*$ be a finite set. Λ is called a partition of Σ if and only if $\Sigma = \cup_{w \in \Lambda} \Sigma_w$ and $\Sigma_w \cap \Sigma_v = \emptyset$ for any $w \neq v \in \Lambda$.

(2) Let Λ_1 and Λ_2 be partitions of Σ. Λ_1 is said to be a refinement of Λ_2 if and only for any $w^1 \in \Lambda_1$, there exists $w^2 \in \Lambda_2$ such that $w^1 \leq w^2$. We write $\Lambda_1 \leq \Lambda_2$ if Λ_1 is a refinement of Λ_2.

For a partition Λ, $\{\Sigma_w\}_{w\in\Lambda}$ is a division of Σ and may be thought of as an approximation of Σ. Note that "$\leq$" is a partial order of the collection of partitions. If $\Lambda_1 \leq \Lambda_2$, then $\{\Sigma_w\}_{w\in\Lambda_1}$ contains finer structure of Σ than $\{\Sigma_w\}_{w\in\Lambda_2}$.

Next we introduce the notion of a scale, which is a monotonically decreasing family of partitions.

DEFINITION 1.1.3 (Scales). A family of partitions of Σ, $\{\Lambda_s\}_{0<s\leq 1}$, is called a scale on Σ if and only if it satisfies (S1) and (S2):
(S1) $\Lambda_1 = W_0$. $\Lambda_{s_1} \leq \Lambda_{s_2}$ for any $0 < s_1 \leq s_2 \leq 1$.
(S2) $\min\{|w||w \in \Lambda_s\} \to +\infty$ as $s \downarrow 0$.

Let $\{\Lambda_s\}_{0<s\leq 1}$ be a scale on Σ. For any $\omega \in \Sigma$ and any $s \in (0,1]$, choose $w \in \Lambda_s$ so that $\omega \in \Sigma_w$, (such a w uniquely exists), and set $U_s(\omega) = \Sigma_w$. Then $\{U_s(\omega)\}_{s\in(0,1]}$ is a system of fundamental neighborhoods of ω. We will think of $U_s(\omega)$ as a "ball" with radius s and center ω even if there may not be a corresponding distance.

In the rest of this section, we will try to understand the basics on scales. First problem is how to describe the structure of a scale.

DEFINITION 1.1.4. Let $\mathcal{S} = \{\Lambda_s\}_{0<s\leq 1}$ be a scale on Σ. For $w \in W_*$. We define

$$R_w(\mathcal{S}) = \{s|s \in (0,1], \text{there exists } w' \in W_* \text{ such that } w < w' \text{ and } w' \in \Lambda_s\},$$
$$C_w(\mathcal{S}) = \{s|s \in (0,1], w \in \Lambda_s\},$$
$$L_w(\mathcal{S}) = \{s|s \in (0,1], \text{there exists } w' \in W_* \text{ such that } w' < w \text{ and } w' \in \Lambda_s\}.$$

For ease of notation, we use R_w, C_w and L_w instead of $R_w(\mathcal{S}), C_w(\mathcal{S})$ and $L_w(\mathcal{S})$ if no confusion can occur. Note that $R_\emptyset = \emptyset$ and that $C_\emptyset$ contains 1.

LEMMA 1.1.5. *Let* $\mathcal{S} = \{\Lambda_s\}_{0<s\leq 1}$ *be a scale on* Σ. *For* $w \in W_*$.
(1) *There exist* $r : W_\# \to (0,1]$ *and* $l : W_* \to (0,1]$ *such that, for any* w, $l(w) \leq r(w)$ *and* $R_w \supseteq (r(w),1]$, $C_w \supseteq (l(w), r(w))$ *and* $L_w \supseteq (0, l(w))$.
(2) *For any* $w \in W_*$ *and any* $i \in S$, $C_{wi} \cup L_{wi} = L_w$. *In particular,* $r(wi) = l(w)$ *and* $l(wi) \leq l(w)$.
(3) $\max\{l(w)|w \in W_m\} \to 0$ *as* $m \to \infty$.

PROOF. (1) Since $1 \in L_\emptyset$, $R_w \neq \emptyset$ for $w \in W_\#$. Also by (S2), $L_w \neq \emptyset$ for any $w \in W_*$. Using (S1), we see that $x < y$ for any $x \in L_w \cup C_w$ and any $y \in R_w$. Therefore the Dedekind theorem implies that there exists $r(w)$ such that $(0, r(w)) \subseteq L_w \cup C_w$ and $(r(w), 1] \subseteq R_w$. In the same manner, we have $l(w)$.
(2) Note that $s \in L_{wi} \cup C_{wi}$ if and only if there exists $w' \in W_*$ such that $w' \leq wi$ and $w' \in \Lambda_s$. This immediately implies that $L_{wi} \cup C_{wi} \subseteq L_w$. Suppose $s \in L_w$. There exists $w' \in \Lambda_s$ such that $wii\ldots \in \Sigma_{w'}$. Since $w'' \notin \Lambda_s$ if $w \leq w''$, it follows that $w' \leq wi$. Therefore $s \in L_{wi} \cup C_{wi}$. The rest of the statement is obvious.
(3) Let $a_m = \max\{l(w)|w \in W_m\}$. Then $a_m \geq a_{m+1}$ for any $m \geq 1$. Set $\alpha = \lim_{m\to\infty} a_m$. Suppose $\alpha > 0$. Choosing $w^m \in W_m$ so that $l(w^m) \geq \alpha$, we see that $\Lambda_{\alpha/2}$ contains $w' \leq w^m$ for any $m \geq 1$. Therefore, $\Lambda_{\alpha/2}$ is an infinite set. This contradiction implies that $\alpha = 0$. □

In general, R_w can be either $(r(w), 1]$ or $[r(w), 1]$. To remove this ambiguity, we introduce the notion of a right continuous scale.

DEFINITION 1.1.6. A scale $\{\Lambda_s\}_{0<s\leq 1}$ on Σ is called right continuous if and only if $R_w = [r(w), 1]$ for any $w \in W_*$.

Lemma 1.1.5 implies that if $\{\Lambda_s\}_{0<s\leq 1}$ is right continuous then $L_w = (0, l(w))$ and $C_w = [l(w), r(w))$.

PROPOSITION 1.1.7. *A scale $\{\Lambda_s\}_{0<s\leq 1}$ on Σ is right continuous if and only if, for any s, there exists $\epsilon > 0$ such that $\Lambda_{s'} = \Lambda_s$ for any $s' \in [s, s+\epsilon)$.*

Right continuous scales are completely determined by $l : W_* \to (0, 1]$, which will be called the gauge function of the scale. See Theorem 1.1.10 for details.

DEFINITION 1.1.8. A function $g : W_* \to (0, 1]$ is called a gauge function on W_* if it satisfies (G1) and (G2):
(G1) $g(wi) \leq g(w)$ for any $w \in W_*$ and any $i \in S$.
(G2) $\max\{g(w)|w \in W_m\} \to 0$ as $m \to \infty$.

The following proposition is immediate by Lemma 1.1.5.

PROPOSITION 1.1.9. *Let $\mathcal{S}$ be a scale on Σ. Then the function $l : W_* \to (0, 1]$ defined in Lemma 1.1.5 is a gauge function on W_*. We call l the gauge function of the scale $\mathcal{S}$.*

Naturally there exists a one to one correspondence between the (right continuous) scales and the gauge functions.

THEOREM 1.1.10. *Let g be a gauge function on W_*. Define $\Lambda_s(g)$ by*

$$\Lambda_s(g) = \{w|w = w_1 \dots w_m \in W_*, g(w_1 \dots w_{m-1}) > s \geq g(w)\} \tag{1.1.8}$$

for any $s \in (0, 1]$. (We regard $g(w_1 \dots w_{m-1})$ as 2 for $w = \emptyset$.) Then $\{\Lambda_s(g)\}_{0<s\leq 1}$ is a right continuous scale on Σ. $\{\Lambda_s(g)\}_{0<s\leq 1}$ is called the scale induced by the gauge function g. Conversely, let $\mathcal{S} = \{\Lambda_s\}_{0<s\leq 1}$ be a right continuous scale on Σ and let l be its gauge function. Then the scale induced by the gauge function l coincides with $\mathcal{S}$.

PROOF. We write $\Lambda_s = \Lambda_s(g)$ for ease of notation. First we show that Λ_s is a finite set for any s. By (G2), there exists $m \geq 1$ such that $s \geq g(w)$ for any $w \in W_m$. Now if $g(v_1 \dots v_{n-1}) > s \geq g(v_1 \dots v_n)$, then $n \leq m$. Therefore $\Lambda_s \subset \cup_{m=0}^m W_m$. Hence Λ_s is a finite set.

Next we show that Λ_s is a partition. Let $\omega = \omega_1\omega_2 \ldots \in \Sigma$. (G2) implies that $g(\omega_1 \dots \omega_m) \to 0$ as $m \to \infty$. Hence there exists a unique m such that $g(\omega_1 \dots \omega_{m-1}) > s \geq g(\omega_1 \dots \omega_m)$. Therefore $\cup_{w \in \Lambda_s} \Sigma_w = \Sigma$. Also the uniqueness of m implies that $\Sigma_{w^1} \cap \Sigma_{w^2} = \emptyset$ if $w^1 \neq w^2 \in \Lambda_s$. So Λ_s is a partition.

To show (S1), since $g(\emptyset) \leq 1$, we have $\Lambda_1 = W_0$. Let $s_1 < s_2$ and let $w = w_1 \dots w_m \in \Lambda_{s_1}$. Then $g(w_1 \dots w_{k-1}) > s_2 \geq g(w_1 \dots w_k)$ for some $k \leq m$. This implies that $w_1 \dots w_k \in \Lambda_{s_2}$. Therefore $\Lambda_{s_1} \leq \Lambda_{s_2}$.

If $s < \min_{w \in W_m} g(w)$, then $\min\{|w||w \in \Lambda_s\} \geq m$. This shows (S2).

Since $R_w = [g(w_1 \dots w_{m-1}), 1]$ for $w = w_1 \dots w_m$, $\{\Lambda_s\}_{0<s\leq 1}$ is right continuous.

Finally, let l be the gauge function of a scale $\mathcal{S} = \{\Lambda_s\}_{0<s\leq 1}$. Then $\Lambda_s = \{w|s \in C_w\} = \Lambda_s(l)$. This completes the proof of the theorem. □

Hereafter, we only consider right continuous scales.

DEFINITION 1.1.11. A right continuous scale $\mathcal{S} = \{\Lambda_s\}_{0<s\le 1}$ on Σ is called elliptic if and only if it satisfies the following two conditions (EL1) and (EL2):
(EL1) $\Lambda_s \cap \Lambda_{\alpha_1 s} = \emptyset$ for any $s \in (0,1]$, where α_1 is independent of s.
(EL2) There exist $\alpha_2 \in (0,1)$ and $n \ge 1$ such that

$$\max\{|v| | v \in W_*, wv \in \Lambda_{\alpha_2 s}\} \le n$$

for any $s \in (0,1]$ and any $w \in \Lambda_s$.

Roughly speaking, a scale is elliptic if the differences between Λ_s and $\Lambda_{\alpha s}$ are uniform with respect to s. This become clearer when we describe (EL1) and (EL2) in terms of gauge functions.

PROPOSITION 1.1.12. *Let $\mathcal{S} = \{\Lambda_s\}_{0<s\le 1}$ be a right continuous scale on Σ and let l be its gauge function.*
(1) *$\mathcal{S}$ satisfies* (EL1) *if and only if there exists $\beta_1 \in (0,1)$ such that $l(wi) \ge \beta_1 l(w)$ for any $w \in W_*$ and any $i \in S$.*
(2) *$\mathcal{S}$ satisfies* (EL2) *if and only if there exist $\beta_2 \in (0,1)$ and $n \ge 1$ such that $l(wv) \le \beta_2 l(w)$ for any $w \in W_*$ and any $v \in W_n$.*

PROOF. (1) First suppose for any $\beta_1 \in (0,1)$ there exist $w \in W_*$ and $i \in S$ such that $l(wi) < \beta_1 l(w)$. In particular, we assume that $\beta_1 < 1/2$. Note that $wi \in \Lambda_s$ for $s \in [l(wi), l(w))$. If $s_1 = l(w)/2$, then $\Lambda_{s_1} \cap \Lambda_{\alpha s_1}$ contains wi for any $\alpha \in [2\beta_1, 1]$. Hence $\mathcal{S}$ does not satisfy (EL1).

Conversely, assume that there exists $\beta_1 \in (0,1)$ such that $l(wi) \ge \beta_1 l(w)$ for any $w \in W_*$ and any $i \in S$. Let $w = w_1 \dots w_m \in \Lambda_s$. Then $l(w_1 \dots w_{m-1}) > t \ge l(w) \ge \beta_1 l(w_1 \dots w_{m-1})$. Therefore $l(w) > \beta_1 s$. This implies that $w \notin \Lambda_{\beta_1 s}$. Hence $\Lambda_s \cap \Lambda_{\beta_1 s}$ is empty for any $s \in (0,1]$.
(2) Assume that there exist $\beta_2 \in (0,1)$ and $n \ge 1$ such that $l(wv) \le \beta_2 l(w)$ for any $w \in W_*$ and any $v \in W_n$. If $w \in \Lambda_s$, then $s \ge l(w)$. Therefore, $\beta_2 s \ge \beta_2 l(w) \ge l(wv)$ for any $v \in W_n$. Hence if $wv' \in \Lambda_{\beta_2 s}$, then $|v|' \le n$. Thus we obtain (EL2) with $\alpha_2 = \beta_2$.

Conversely, suppose that, for any $\beta \in (0,1)$ and any $k \ge 1$, there exist $w \in W_*$ and $v \in W_k$ such that $\beta l(w) \le l(wv)$. Let $s = l(w)$. Here, if necessary, replacing $w = w_1 \dots w_i$ by $w = w_1 \dots w_j$ for some $0 \le j \le i$, we may assume that $w \in \Lambda_s$. (Then, in general, $|v| \ge k$.) Now choose $\beta > \alpha_2$ and $k \ge n$. Then $\alpha_2 s < \beta s = \beta l(w) \le l(wv)$. Therefore there exists $v' \in W_*$ such that $wvv' \in \alpha_2 s$. By (EL2), $|vv'| \le n$. This contradicts to the fact that $|v| \ge k$. □

The following fact will be used later in many places.

LEMMA 1.1.13. *Let $\mathcal{S} = \{\Lambda_s\}_{0<s\le 1}$ be a scale on Σ satisfying* (EL1) *and let l be its gauge function on W_*. Then there exists a constant $c > 0$ such that $l(w) \le s \le cl(w)$ for any $s \in (0,1]$ and any $w \in \Lambda_s$.*

PROOF. Let $w = w_1 \dots w_m \in \Lambda_s$. Then $l(w_1 \dots w_{m-1}) > s \ge l(w)$. By Proposition 1.1.12-(1), $l(w) \ge \beta_1 l(w_1 \dots w_{m-1})$. Therefore, if $c = 1/\beta_1$, then $cl(w) \ge s \ge l(w)$. □

Next we define a multiplication of two scales and a power of a scale.

DEFINITION 1.1.14. (1) For $i = 1, 2$, let $\mathcal{S}_j$ be a scale on Σ and let l_j be its gauge function. Then we use $\mathcal{S}_1 \cdot \mathcal{S}_2$ to denote the sale induced by the gauge function $l_1 l_2$.
(2) Let $\mathcal{S}$ be a scale on Σ. Then for $\alpha > 0$, the scale induced by the gauge function l^α is denoted by $\mathcal{S}^\alpha$.

If $\mathcal{S} = \{\Lambda_s\}_{0<s\leq 1}$, then $\mathcal{S}^\alpha = \{\Lambda_{s^{1/\alpha}}\}_{0<s\leq 1}$.

LEMMA 1.1.15. (1) *If* $\mathcal{S}_1$ *and* $\mathcal{S}_2$ *are elliptic scales on* Σ, *then* $\mathcal{S}_1 \cdot \mathcal{S}_2$ *is elliptic.*
(2) *Let* $\mathcal{S}$ *be a scale on* Σ *and let* $\alpha > 0$. *Then* $\mathcal{S}$ *is elliptic if and only if* $\mathcal{S}^\alpha$ *is elliptic.*

Finally we introduce an important class of scales.

DEFINITION 1.1.16. Let $\mathbf{a} = (a_i)_{i\in S} \in (0,1)^S$. Define $g_{\mathbf{a}} : W_* \to (0,1]$ by $g_{\mathbf{a}}(w) = a_w = a_{w_1}a_{w_2}\ldots a_{w_m}$ for $w = w_1\ldots w_m \in W_*$. $g_{\mathbf{a}}$ is called the self-similar gauge function on W_* with weight $\mathbf{a}$. Also the scale induced by $g_{\mathbf{a}}$ is called the self-similar scale with weight $\mathbf{a}$ and is denoted by $\mathcal{S}(\mathbf{a})$. We also write $\Lambda_s(g_{\mathbf{a}}) = \Lambda_s(\mathbf{a})$. We use $\mathfrak{S}(\Sigma)$ to denote the collection of self-similar scales on Σ.

We often identify $\mathfrak{S}(\Sigma)$ with $(0,1)^S$ through the natural correspondence $\mathbf{a} \to \mathcal{S}(\mathbf{a})$. Note that a self-similar scale is elliptic.

1.2. Self-similar structures and measures

The notion of self-similar structure is a purely topological formulation of self-similar sets.

DEFINITION 1.2.1. (1) Let K be a compact metrizable topological space and let S be a finite set. Also, let F_i, for $i \in S$, be a continuous injection from K to itself. Then, $(K, S, \{F_i\}_{i\in S})$ is called a self-similar structure if there exists a continuous surjection $\pi : \Sigma \to K$ such that $F_i \circ \pi = \pi \circ \sigma_i$ for every $i \in S$.
(2) Let $\mathcal{L} = (K, S, \{F_i\}_{i\in S})$ be a self-similar structure. Define the critical set $\mathcal{C}_{\mathcal{L}}$ and the post critical set $\mathcal{P}_{\mathcal{L}}$ by $\mathcal{C}_{\mathcal{L}} = \pi^{-1}(\cup_{i\neq j\in S}(F_i(K)\cap F_j(K)))$ and $\mathcal{P}_{\mathcal{L}} = \cup_{n\geq 1}\sigma^n(\mathcal{C}_{\mathcal{L}})$. Also define $V_0 = \pi(\mathcal{P}_{\mathcal{L}})$.
(3) A self-similar structure $\mathcal{L} = (K, S, \{F_i\}_{i\in S})$ is said to be strongly finite if and only if $\sup_{x\in K} \#(\pi^{-1}(x)) < +\infty$, where $\#(A)$ is the number of elements of a set A.

NOTATION. Let $\mathcal{L} = (K, S, \{F_i\}_{i\in S})$ be a self-similar structure. For $w = w_1\ldots w_m \in W_*$, we define $F_w = F_{w_1} \circ \ldots \circ F_{w_m}$ and $K_w = F_w(K)$. In particular, if $w = \emptyset \in W_0$, then F_w is thought of as the identity map of K and $K_w = K$.

If $(K, S, \{F_i\}_{i\in S})$ is a self-similar structure, then

$$K = \bigcup_{i\in S} F_i(K).$$

In other words, K is the self-similar set with respect to maps $\{F_i\}_{i\in S}$. The set V_0 is a kind of "boundary" of K. Indeed, for any $w, v \in W_*$ with $\Sigma_w \cap \Sigma_v = \emptyset$, $K_w \cap K_v = F_w(V_0) \cap F_v(V_0)$. Moreover, $V_0 = \emptyset$ if and only if π is bijective and K is identified with Σ. V_0 is thought of as a characteristic of "complexity" of the self-similar structure.

Throughout this section, we fix a self-similar structure $\mathcal{L} = (K, S, \{F_i\}_{i\in S})$.

THEOREM 1.2.2. *$K \neq \overline{V}_0$ if and only if* $\mathrm{int}(\overline{V}_0) = \emptyset$.

PROOF. Assume that $\mathrm{int}(\overline{V}_0) \neq \emptyset$. Then $\overline{V}_0 \supseteq K_w$ for some $w \in W_*$. Let $|w| = m$. Let $x \in K_w$. If $x \in K_w \cap (\cup_{v \in W_m, v \neq w} K_v)$, then $x \in F_w(V_0)$ by [**28**, Proposition 1.3.5]. Hence $(F_w)^{-1}(x) \in \overline{V}_0$. Next we assume that $x \notin K_w \cap (\cup_{v \in W_m, v \neq w} K_v)$. This assumption is equivalent to that $\pi^{-1}(x) \subset \Sigma_w$. Since $x \in \overline{V}_0$, there exist $\omega(1), \omega(2), \ldots \in \mathcal{P}$ such that $\pi(\omega(i)) \to x$ as $i \to \infty$. Choosing a subsequence, we may assume that there exists $\omega \in \Sigma$ such that $\omega(i) \to \omega$ as $i \to \infty$. The continuity of π implies that $\pi(\omega) = x$. Hence $\omega \in \Sigma_w$. Now $\sigma^m \omega(i) \to \sigma^m \omega$ as $i \to \infty$ and $(F_w)^{-1}(x) = \pi(\sigma^m \omega)$. Since $\sigma^m \omega(i) \in \mathcal{P}$, it follows that $(F_w)^{-1}(x) \in \overline{V}_0$. Thus we see that $(F_w)^{-1}(x) \in \overline{V}_0$ for any $x \in K_w$. This immediately implies that $\overline{V}_0 = K$. The converse is obvious. □

Next we introduce a class of non-degenerate measures on a self-similar structure.

DEFINITION 1.2.3. $\mathcal{M}(K)$ is defined to be the collection of Borel regular measures on K satisfying the following conditions (M1), (M2) and (M3):
(M1) μ is a finite Borel regular measure on K.
(M2) For any $w \in W_*$, $\mu(K_w) > 0$ and $\mu(F_w(V_0)) = 0$.
(M3) $\mu(\{x\}) = 0$ for any $x \in K$.
Also $\mathcal{M}_1(K) = \{\mu | \mu \in \mathcal{M}(K), \mu(K) = 1\}$.

THEOREM 1.2.4. *Assume that $K \neq \overline{V}_0$. Let μ be a finite Borel regular measure on K with $\mu(K) > 0$. Then $\mu \in \mathcal{M}(K)$ if the following condition* (ELm) *holds:*
(ELm) *there exists $\gamma > 0$ such that $\mu(K_{wi}) \geq \gamma \mu(K_w)$ for any $w \in W_*$ and any $i \in S$.*

DEFINITION 1.2.5. A finite Borel regular measure μ on K is called an elliptic measure if and only if it satisfies (ELm).

REMARK. In [**28**, Section 3.4], a Borel regular measure μ satisfying (ELm) is called a γ-elliptic measure. By the proof of [**28**, Lemma 3.4.1], it follows that if μ is elliptic then there exists $\delta \in (0,1)$ and $m \geq 1$ such that $\mu(K_{wv}) \leq \delta \mu(K_w)$ for any $w \in W_*$ and any $v \in W_m$.

PROOF. (M1) is immediate. Since $K \neq \overline{V}_0$, there exist $k \in \mathbb{N}$ and $v \in W_k$ such that $K_v \cap \overline{V}_0 = \emptyset$. Since μ is a finite Borel regular measure, for any $\epsilon > 0$, we find an open set O which satisfies $F_w(V_0) \subset O$ and $\mu(O) \leq \mu(F_w(V_0)) + \epsilon$. Set $Q = \{\tau | \tau \in W_*, K_\tau \subset O |\tau| \geq |w|\}$. As O is open, $O = \cup_{\tau \in Q} K_\tau$. Define $Q_* = \{\tau | \tau \in Q, \text{there exists no } \tau' \in Q \text{ such that } \tau < \tau'\}$. Then $\{K_{\tau v}\}_{\tau \in Q_*}$ is mutually disjoint. Also, for any $\tau \in Q_*$, since $K_{\tau v} \cap V_{|w|} = \emptyset$, we see that $F_w(V_0) \cap K_{\tau v} = \emptyset$. Therefore, by the fact that μ is elliptic,

$$\mu(F_w(V_0)) \leq \mu(O \backslash \cup_{\tau \in Q_*} K_{\tau v}) = \mu(O) - \sum_{\tau \in Q_*} \mu(K_{\tau v})$$

$$\leq \mu(O) - \gamma^k \sum_{\tau \in Q_*} \mu(K_\tau) \leq (1 - \gamma^k)\mu(O) \leq (1 - \gamma^k)(\mu(F_w(V_0)) + \epsilon)$$

Since this holds for any $\epsilon > 0$, it follows that $\mu(F_w(V_0)) = 0$. Thus we obtain (M2).

To show (M3), let $x = \pi(\omega)$ for $\omega \in \Sigma$. By the above remark, we see that $\mu(K_{\omega_1 \ldots \omega_{mn}}) \leq \delta^n$ for any n. Therefore, $\mu(\{x\}) = \mu(\cap_{n \geq 1} K_{\omega_1 \ldots \omega_{mn}}) = 0$. □

A immediate example of a elliptic measure is a self-similar measure.

DEFINITION 1.2.6. Let $(\mu_i)_{i\in S} \in (0,1)^S$ satisfy $\sum_{i\in S}\mu_i = 1$. A Borel regular probability measure μ is called a self-similar measure with weight $(\mu_i)_{i\in S}$ if and only if

$$\mu(A) = \sum_{i\in S}\mu_i\mu((F_i)^{-1}(A)) \tag{1.2.1}$$

for any Borel set A.

It is known that, for any weight $(\mu_i)_{i\in S}$, there exists a unique Borel regular probability measure on k that satisfies (1.2.1). See [**28**, Section 1.4]. In our case, we have the following theorem.

THEOREM 1.2.7. *Assume that $K \neq \overline{V}_0$. Let μ be a self-similar measure with weight $(\mu_i)_{i\in S} \in (0,1)^S$ with $\sum_{i\in S}\mu_i = 1$. Then, $\mu(K_w) = \mu_w$ for any $w \in W_*$, where $\mu_w = \mu_{w_1}\cdots\mu_{w_m}$ for $w = w_1\dots w_m$. In particular, μ is elliptic and $\mu \in \mathcal{M}_1(K)$.*

PROOF. Let $O = K\backslash\overline{V}_0$. For any $w \in W_m$, $K_w = F_w(O) \cup F_w(\overline{V}_0)$ and $F_w(O) \cap F_w(\overline{V}_0) = \emptyset$. By [**28**, Proposition 1.3.5], $F_w(O) \cap K_v = \emptyset$ for any $v \in W_m$ with $w \neq v$. Therefore, $F_w(O)$ is open. Moreover, since $V_0 \subseteq V_m = \cup_{v\in W_m}F_w(V_0)$, it follows that $F_w(O)\cap V_0 = \emptyset$. This implies that $F_w(O)\cap\overline{V}_0 = \emptyset$. Hence $F_w(O) \subseteq O$.

On the other hand, by (1.2.1),

$$\mu(F_w(O)) = \cup_{v\in W_m}\mu_v\mu((F_v)^{-1}F_w(O)) = \mu_w\mu(O).$$

Therefore, if $O_m = \cup_{w\in W_m}F_w(O)$, we obtain $\mu(O_m) = \sum_{w\in W_m}\mu(F_w(O)) = \mu(O)$. Note that $O_m \subseteq O$. For sufficiently large m, there exists $w \in W_m$ such that $K_w \subset O$. Since $F_w(\overline{V}_0) \cap O_m = \emptyset$, $F_w(\overline{V}_0) \in O\backslash O_m$. Therefore $\mu(F_w(\overline{V}_0)) = 0$. By (1.2.1), $\mu(\overline{V}_0) = 0$ and therefore $\mu(O) = \mu(O_k) = 0$ for any $k \geq 1$. This implies that $\mu(\cup_{w\in W_k}F_w(\overline{V}_0)) = 0$ for any $k \geq 1$. Hence for any $w \in W_*$, $\mu(K_w) = \mu(F_w(O)) = \mu_w\mu(O) = \mu_w$. This immediately shows that μ is elliptic. Now by Theorem 1.2.4, we verify $\mu \in \mathcal{M}_1(K)$. □

REMARK. In the above proof, it was shown that if $K \neq \overline{V}_0$, then $\mathcal{L}$ satisfy an "intrinsic" open set condition: there exists a nonempty intrinsic open subset $O \subset K$ (i.e. O is open with respect to the topology of K) such that $F_i(O) \subseteq O$ and $F_i(O) \cap F_j(O) = \emptyset$ for any $i \neq j \in S$.

REMARK. We conjecture that the converse of the above theorem is true: if every self-similar measure μ belongs to $\mathcal{M}_1(K)$ (and hence $\mu(K_w) = \mu_w$ for any $w \in W_*$, where $(\mu_i)_{i\in S}$ is the weight of μ), then $K \neq \overline{V}_0$.

We may define a natural gauge function associated with a measure as follows.

PROPOSITION 1.2.8. *Let $\mu \in \mathcal{M}^1(K)$. Define $g_\mu : W_* \to (0,1]$ by $g_\mu(w) = \mu(K_w)$. Then g_μ is a gauge function on W_*.*

PROOF. (G1) is immediate. To prove (G2), assume that there exists $\alpha > 0$ such that $\max\{\mu(K_w)|w \in W_m\} \geq 2\alpha$ for any $m \geq 1$. Then $A = \{w|w \in W_*, \mu(K_w) \geq \alpha\}$ is an infinite set. Let $T = \{w|w \in A, \{v|v \in A, v < w\} = \emptyset\}$. If $w^1, \cdots, w^k \in T$ and $w^i \neq w^j$ for any $i \neq j$, then (M2) implies that $\mu(\cup_{i=1}^k K_{w^i}) = \sum_{i=1}^k \mu(K_{w^i}) \geq k\alpha$. Hence $k \leq 1/\alpha$. So T is a finite set. Set $M = \max_{w\in T}|w|$ and choose $w \in A$ with $|w| > M$. Then there exists a sequence $\{w(i)\}_{i=1,2,\dots} \subset A$ such that

$w(1) = w$ and $w(1) \geq w(2) \geq w(3) \geq \ldots$. Let $x = \pi(\omega)$, where $\pi : \Sigma \to K$ is given in Definition 1.2.1 and $\omega \in \Sigma$ is the unique infinite sequence contained in $\cap_{i=1,2,\ldots}\Sigma_{w(i)}$. Then $\mu(\{x\}) = \lim_{i\to\infty} \mu(K_{w(i)}) \geq \alpha$. This contradicts to (M3). Hence we have verified (G2). □

DEFINITION 1.2.9. Let $\mu \in \mathcal{M}(K)$ and denote $\bar{\mu} = \mu/\mu(K)$. Then $g_{\bar{\mu}}$ defined in Proposition 1.2.8 is called the gauge function on W_* induced by the measure μ. If no confusion can occur, we use μ to denote $g_{\bar{\mu}}$ and write $\Lambda_s(\mu) = \Lambda_s(g_{\bar{\mu}})$. $\{\Lambda_s(\mu)\}_{0<s\leq 1}$ is called a scale on Σ induced by the measure μ.

The following two facts are immediate from the definition.

PROPOSITION 1.2.10. *Let μ be a self-similar measure. Then the scale induced by μ is an self-similar scale with the same weight as μ.*

PROPOSITION 1.2.11. *Let $\mu \in \mathcal{M}(K)$. Then the scale induced by μ is elliptic if and only if μ is elliptic.*

1.3. Volume doubling property

In this section, S is a finite set, $\mathcal{L} = (K, S, \{F_i\}_{i\in S})$ is a self-similar structure with $K \neq \overline{V}_0$. Also $\mathcal{S} = \{\Lambda_s\}_{0<s\leq 1}$ is a right-continuous scale on Σ and μ is always assumed to be a Borel regular finite measure on K which belongs to $\mathcal{M}(K)$. We will introduce a system of neighborhoods $\{U_s^{(n)}(x)\}$ of $x \in K$ associated with the scale Σ and consider the counterpart of "volume doubling measures" on K.

Ordinarily, if (X, d) is a metric space and μ is a Borel measure on (X, d), then μ is said to have the volume doubling property (or μ is volume doubling in short) if $\mu(B_{2r}(x)) \leq C\mu(B_r(x))$ for any $x \in X$ and $r > 0$, where $B_r(x)$ is a ball $B_r(x) = \{y|d(x,y) < r\}$ and C is a constant which is independent of x and r. We will think of $U_s^{(n)}(x)$ as a ball and introduce the notion corresponding to the volume doubling property. The main goal of this section, which is Theorem 1.3.5, is to establish conditions which are equivalent to the volume doubling property in our framework.

To start with, we associate subsets of words with those of self-similar sets.

DEFINITION 1.3.1. Let $\Gamma \subseteq W_*$ and let $A \subseteq K$.
(1) $W(\Gamma, A) = \{w|w \in \Gamma, K_w \cap A \neq \emptyset\}$.
(2) $K(\Gamma) = \cup_{w\in\Gamma}K_w$.
(3) Define $W^{(n)}(\Gamma, A)$ and $K^{(n)}(\Gamma, A)$ by $W^{(0)}(\Gamma, A) = W(\Gamma, A)$, $K^{(n)}(\Gamma, A) = K(W^{(n)}(\Gamma, A))$ and $W^{(n+1)}(\Gamma, A) = W(\Gamma, K^{(n)}(\Gamma, A))$ for $n = 0, 1, \ldots$.
(4) We use ∂A be the topological boundary of A, i.e. $\partial A = \overline{A^c} \cap \overline{A}$.

Under a scale $\mathcal{S} = \{\Lambda_s\}_{0<s\leq 1}$, the "radius" of K_w is thought of as s if $w \in \Lambda_s$. In this way we may define a ball of radius s with respect to a scale in the following way.

DEFINITION 1.3.2. Let $\mathcal{S} = \{\Lambda_s\}_{0<s\leq 1}$ be a scale on Σ. For $x \in K$, we write $\Lambda_{s,x}^n = W^{(n)}(\Lambda_s, \{x\})$ and $U_s^{(n)}(x) = K^{(n)}(\Lambda_s, \{x\})$ for $n \geq 0$. In particular, we use $\Lambda_{s,x} = \Lambda_{s,x}^0, K_s(x) = U_s^{(0)}(x)$ and $U_s(x) = U_s^{(1)}(x)$. Also set $\Lambda_{s,w} = W(\Lambda_s, K_w)$ for $w \in W_*$.

$U_s^{(n)}(x)$ is a neighborhood of x for any n. In the case $n = 0$, however, $K_s(x) = U_s^{(0)}(x)$ is not a good ball with center x since x may be very close to the boundary of $K_s(x)$ i.e. $\partial K_s(x)$. (Note that if $x \in K_w \backslash F_w(V_0)$ and $w \in \Lambda_s$, then $K_s(x) = K_w$.) This will make a crucial difference between the role of $\{U_s^{(n)}(x)\}_{x \in K, s>0}$ for $n = 0$ and that for $n \geq 1$.

DEFINITION 1.3.3. Let $\mathcal{S} = \{\Lambda_s\}_{0<s\leq 1}$ be a scale on Σ and let $\mu \in \mathcal{M}(K)$. For $n \geq 0$, we define a property $(\mathrm{VD})_\mathrm{n}$ on $(\mathcal{S}, \mu)$ as follows.

$(\mathrm{VD})_\mathrm{n}$ There exist $\alpha \in (0,1)$ and $c_V > 0$ such that $\mu(U_s^{(n)}(x)) \leq c_V \mu(U_{\alpha s}^{(n)}(x))$ for any $s \in (0,1]$ and any $x \in K$.

If $(\mathcal{S}, \mu)$ satisfy $(\mathrm{VD})_\mathrm{n}$ for some $n \geq 1$, we say that μ has the volume doubling property (or (VD) for short) with respect to $\mathcal{S}$.

If $\mathcal{S}$ satisfies (EL1), then $(\mathrm{VD})_\mathrm{n}$ will be shown to be equivalent to $(\mathrm{VD})_1$ for any $n \geq 1$ in Theorem 1.3.11. Therefore, μ has (VD) with respect to $\mathcal{S}$ if and only if $(\mathrm{VD})_\mathrm{n}$ holds for all $n \geq 1$.

We introduce several notions to describe the conditions which is equivalent to the volume doubling property.

DEFINITION 1.3.4. (1) Let $\varphi : W_* \to [0, \infty)$. We say that φ is gentle with respect to $(\mathcal{L}, \mathcal{S})$ if and only if it satisfies the following condition (GE):

(GE) There exists $c_G > 0$ such that $\varphi(w) \leq c_G \varphi(v)$ for any $s \in (0,1]$ and any $w, v \in \Lambda_s$ with $K_w \cap K_v \neq \emptyset$.

μ is said to be gentle with respect to $\mathcal{S}$ if and only if φ_μ is gentle with respect to $(\mathcal{L}, \mathcal{S})$, where φ_μ is defined by $\varphi(w) = \mu(K_w)$.

(2) $\mathcal{S}$ is said to be locally finite with respect to $\mathcal{L}$ if and only if it satisfies the following condition (LF):

(LF) $\sup\{\#(\Lambda_{s,x}^1) | s \in (0,1], x \in K\} < +\infty$,

(3) Let $n \in \mathbb{N}$. We define properties $(\mathrm{A})_\mathrm{n}$ on $(\mathcal{S}, \mu)$ as follows.

$(\mathrm{A})_\mathrm{n}$ There exists $c_A > 0$ such that $\mu(U_s^{(n)}(x)) \leq c_A \mu(K_s(x))$ for any $s \in (0,1]$ and any $x \in K$.

Note that the notion of "gentle measure" concerns both a scale and a measure while the condition (LF) is determined solely by a scale.

THEOREM 1.3.5. *Assume that* $\mathcal{S} = \{\Lambda_s\}_{0<s\leq 1}$ *is elliptic. Let* $n \geq 1$. *Then the following three conditions are equivalent.*

(1) $\mathcal{S}$ *is locally finite and* μ *is elliptic and gentle with respect to* $\mathcal{S}$. *In short,* (LF) $\wedge$ (ELm) $\wedge$ (GE).

(2) $(\mathcal{S}, \mu)$ *has properties* $(\mathrm{A})_\mathrm{n}$ *and* $(\mathrm{VD})_0$. *In short,* $(\mathrm{A})_\mathrm{n} \wedge (\mathrm{VD})_0$.

(3) $(\mathcal{S}, \mu)$ *satisfies* $(\mathrm{VD})_\mathrm{n}$.

In particular, $(\mathrm{VD})_\mathrm{n}$ *is equivalent to* $(\mathrm{VD})_1$ *for any* $n \in \mathbb{N}$ *and* (VD) $\Leftrightarrow$ (LF) $\wedge$ (ELm) $\wedge$ (GE).

REMARK. We will show stronger statement on the equivalence between (1) and (2). In fact, by Theorems 1.3.8 and 1.3.10, (LF) $\wedge$ (GE) $\Leftrightarrow$ $(\mathrm{A})_\mathrm{n}$ and (ELm) $\Leftrightarrow$ $(\mathrm{VD})_0$.

The main purpose of the rest of this section is to prove the above theorem. First we examine the condition (LF).

LEMMA 1.3.6. *The following three conditions are equivalent:*
(1) $\mathcal{S}$ *is locally finite with respect to* $\mathcal{L}$.
(2) $\sup\{\#(\Lambda_{s,w})|s \in (0,1], w \in \Lambda_s\} < +\infty$.
(3) $\sup\{\#(\Lambda^n_{s,x})|s \in (0,1], x \in K\} < +\infty$ *for any* $n \geq 1$.
Moreover, if $\mathcal{S}$ *is locally finite with respect to* $\mathcal{L}$*, then* $\mathcal{L}$ *is strongly finite.*

PROOF. (1) $\Rightarrow$ (2): Let $s \in (0,1]$ and let $w \in \Lambda_s$. Choose $x \in K_w$. Then $\Lambda_{s,w} \subseteq \Lambda^1_{s,x}$. This immediately implies (2).
(2) $\Rightarrow$ (3): Set $M = \sup\{\#(\Lambda_{s,w})|s \in (0,1], w \in \Lambda_s\}$. First we show that (2) implies that $\mathcal{L}$ is strongly finite. For any $x \in K$, if $\#(\pi^{-1}(x)) \geq m$, then $\#(\Lambda_{s,x}) \geq m$ for sufficiently small s. Choose $w \in \Lambda_{s,x}$, then $\#(\Lambda_{s,w}) \geq m$. Therefore (2) implies that $\#(\pi^{-1}(x)) \leq M$ and $\#(\Lambda_{s,x}) \leq M$. Note $\Lambda^n_{s,x} = \cup_{w \in \Lambda^{n-1}_{s,x}} \Lambda_{s,w}$. Hence $\#(\Lambda^n_{s,x}) \leq M^{n+1}$ for any $n \geq 1$.
(3) $\Rightarrow$ (1): Obvious. $\square$

If μ satisfy $(\mathrm{VD})_\mathrm{n}$, then $\mu(U^{(n)}_s(x)) \leq (c_V)^m \mu(U^{(n)}_{\alpha^m s}(x))$ for any $m \geq 1$. This fact lead us to the following proposition.

PROPOSITION 1.3.7. *Let* $n \geq 0$. $(\mathrm{VD})_\mathrm{n}$ *is equivalent to the following stronger condition:*
For any $\alpha \in (0,1)$*, there exists* $c > 0$ *such that* $\mu(U^{(n)}_s(x)) \leq c\mu(U^{(n)}_{\alpha s}(x))$ *for any* $s \in (0,1]$ *and any* $x \in K$.

Now we give the first piece of a proof of Theorem 1.3.5.

THEOREM 1.3.8. *Let* $n \in \mathbb{N}$. *Then* μ *is gentle with respect to the scale* $\mathcal{S}$ *and* $\mathcal{S}$ *is locally finite if and only if the property* $(\mathrm{A})_\mathrm{n}$ *is satisfied. In short,* $(\mathrm{GE}) \wedge (\mathrm{LF}) \Leftrightarrow (\mathrm{A})_\mathrm{n}$. *In particular,* $(\mathrm{A})_\mathrm{n}$ *and* $(\mathrm{A})_\mathrm{m}$ *are equivalent to each other for any* $n, m \in \mathbb{N}$.

PROOF. $(\mathrm{GE}) \wedge (\mathrm{LF}) \Rightarrow (\mathrm{A})_\mathrm{n}$: For any $w \in \Lambda^n_{s,x}$, there exist $w^0, w^1, \ldots, w^n \in \Lambda_s$ such that $w^0 \in \Lambda_{s,x}$, $w^n = w$ and $K_{w^{j-1}} \cap K_{w^j} \neq \emptyset$ for $j = 1, \ldots, n$. Hence by (GE),

$$\mu(U^{(n)}_s(x)) = \sum_{w \in \Lambda^n_{s,x}} \mu(K_w) \leq$$
$$(c_G)^n \#(\Lambda^n_{s,x}) \max_{w \in \Lambda_{s,x}} \mu(K_w) \leq (c_G)^n \#(\Lambda^n_{s,x}) \mu(K_s(x)).$$

Therefore by Lemma 1.3.6, (LF) implies $(\mathrm{A})_\mathrm{n}$.
$(\mathrm{A})_\mathrm{n} \Rightarrow$ (GE): Note that $(\mathrm{A})_\mathrm{n}$ implies $(\mathrm{A})_1$. Let $w, v \in \Lambda_s$ satisfy $K_w \cap K_v \neq \emptyset$. Since $K \backslash \overline{V}_0 \neq \emptyset$, $K_w \backslash F_w(V_0) \neq \emptyset$. Choose $x \in K_w \backslash F_w(V_0)$. Then $K_s(x) = K_w$. By $(\mathrm{A})_1$, $c_A \mu(K_w) = c_A \mu(K_s(x)) \geq \mu(U_s(x)) \geq \mu(K_v)$.
$(\mathrm{A})_\mathrm{n} \wedge (\mathrm{GE}) \Rightarrow (\mathrm{LF})$: Let $w \in \Lambda_s$. Choosing $x \in K_w \backslash F_w(V_0)$, we see that $K_s(x) = K_w$ and $\Lambda^1_{s,x} = \Lambda_{s,w}$. By $(\mathrm{A})_1$ and (GE),

$$c_A \mu(K_s(x)) \geq \mu(U_s(x)) \geq c_G \#(\Lambda^1_{s,x}) \mu(K_w) = c_G \#(\Lambda^1_{s,x}) \mu(K_s(x)).$$

Dividing this by $\mu(K_s(x))$, we obtain (LF). $\square$

The second piece of Theorem 1.3.5 is the equivalence between $(\mathrm{VD})_0$ and (ELm). To give an exact statement, we need a definition.

DEFINITION 1.3.9. $(\mathcal{L}, \mathcal{S}, \mu)$ is said to have the property (ELmg) if and only if the following condition is satisfied:
There exist $\alpha \in (0,1)$ and $c_E > 0$ such that $\mu(K_{wv}) \geq c_E \mu(K_w)$ for any $s \in (0,1]$, any $w \in \Lambda_s$ and any $v \in W_*$ with $wv \in \Lambda_{\alpha s}$.

REMARK. If (ELmg) is satisfied, then, for any $\beta \in (0,1)$, there exists $c > 0$ such that $\mu(K_{wv}) \geq c\mu(K_w)$ for any $s \in (0,1)$, any $w \in \Lambda_s$ and any $v \in W_*$ with $wv \in \Lambda_{\beta s}$. Indeed, for any $k > 0$, $\mu(K_{wv}) \geq \alpha^k \mu(K_w)$ for any $s \in \Lambda_s$, any $w \in \Lambda_s$ and any $v \in W_*$ with $wv \in \Lambda_{\alpha^k s}$.

THEOREM 1.3.10. $(\mathrm{VD})_0 \Leftrightarrow$ (ELmg) $\Leftrightarrow$ (ELm) $\wedge$ (EL2). *In particular, if* $\mathcal{S}$ *is elliptic, then* $(\mathrm{VD})_0$ *is equivalent to* (ELm).

PROOF. $(\mathrm{VD})_0 \Rightarrow$ (ELmg): Let $w \in \Lambda_s$ and let $wv \in \Lambda_{\alpha s}$. Choose $x \in K_{wv} \backslash F_{wv}(V_0)$. Then $K_{\alpha s}(x) = K_{wv}$. Hence by $(\mathrm{VD})_0$, $c_B \mu(K_{wv}) = c_B \mu(K_{\alpha s}(x)) \geq \mu(K_s(x))$. Since $K_s(x) \supseteq K_w$, $c_B \mu(K_{wv}) \geq \mu(K_w)$. This shows (ELmg).
(ELmg) $\Rightarrow$ (ELm) $\wedge$ (EL2): Let g be the gauge function of the scale $\mathcal{S}$. For any $w = w_1 \ldots w_m \in W_*$, there exists $n \geq 0$ such that $g(w) = g(w_1 \ldots w_{m-n}) < g(w_1 \ldots w_{m-n-1})$. Set $w' = w_1 \ldots w_{m-n}$. Let $s = g(w)$. Note that $w' \in \Lambda_s$. For any $i \in S$, we can find $v \in W_*$ which satisfies $wiv \in \Lambda_{\alpha s}$. By (ELmg),

$$\mu(K_{wi}) \geq \mu(K_{wiv}) \geq c\mu(K_{w'}) \geq c\mu(K_w).$$

This shows (ELm). Recall the remark after Definition 1.2.5. Under (ELm), there exist $\delta \in (0,1)$ and $k \geq 1$ such that $\mu(K_{wv}) \leq \delta \mu(K_w)$ for any $w \in W_*$ and any $v \in W_k$. Therefore (ELmg) implies

$$c_E \mu(K_w) \leq \mu(K_{wv}) \leq \delta^{[|v|/k]} \mu(K_w).$$

for any $s \in (0,1]$, any $w \in \Lambda_s$ and any $v \in W_*$ with $wv \in \Lambda_{\alpha s}$, where $[x]$ is the integral part of x. Dividing this by $\mu(K_w)$, we have uniform upper estimate of $|v|$. This shows (EL2).
(ELm) $\wedge$ (EL2) $\Rightarrow (\mathrm{VD})_0$: Let $x \in K$. For any $w \in \Lambda_{s,x}$, we may choose $v(w) \in W_*$ so that $wv(w) \in \Lambda_{\alpha s}$. By (EL2), we have $|v(w)| \leq m$, where m is independent of x, s and w. Using (ELm), we obtain $\mu(K_{wv(w)}) \geq c^m \mu(K_w)$. Hence,

$$\mu(K_{\alpha s}(x)) \geq \sum_{w \in \Lambda_{s,x}} \mu(K_{wv(w)}) \geq \sum_{w \in \Lambda_{s,x}} c^m \mu(K_w) = c^m \mu(K_s(x)).$$

Therefore we have $(\mathrm{VD})_0$. □

The next theorem is a generalized version of Theorem 1.3.5.

THEOREM 1.3.11. *Let* $n \geq 1$. *Assume that* $\mathcal{S}$ *satisfies* (EL1). *Then the following three conditions are equivalent:*
(1) $\mathcal{S}$ *is locally finite,* μ *is gentle with respect to* $\mathcal{S}$ *and satisfy* $(\mathrm{VD})_0$. *In short,* (LF) $\wedge$ (GE) $\wedge$ $(\mathrm{VD})_0$.
(2) $(\mathcal{S}, \mu)$ *has properties* $(\mathrm{A})_\mathrm{n}$ *and* $(\mathrm{VD})_0$.
(3) $(\mathcal{S}, \mu)$ *satisfies* $(\mathrm{VD})_\mathrm{n}$.
In particular, $(\mathrm{VD})_\mathrm{n}$ *is equivalent to* $(\mathrm{VD})_1$ *for any* $n \in \mathbb{N}$ *and* (VD) $\Leftrightarrow$ (LF) $\wedge$ (GE) $\wedge$ $(\mathrm{VD})_0$.

To prove the above theorem, we need the following lemma.

LEMMA 1.3.12. *Let* $n \in \mathbb{N}$. *If* $\mathcal{S}$ *satisfy* (EL1), *then there exist* $\alpha \in (0,1)$ *and* $z \in K$ *such that* $K_w \supseteq U_{\alpha s}^{(n)}(F_w(z))$ *for any* $s \in (0,1]$ *and any* $w \in \Lambda_s$.

PROOF. Choose $z \in K\backslash\overline{V}_0$. Since $K\backslash\overline{V}_0$ is open, there exists $\beta \in (0,1)$ such that $U_\beta^{(n)}(z) \subseteq K\backslash\overline{V}_0$. Set $m = \max_{v\in\Lambda_\beta}|v|$. Let $w \in \Lambda_s$. Note that (EL1) implies that $|v| \geq m$ for any $wv \in \Lambda_{(\alpha_1)^m s}$, where α_1 is the constant appeared in Definition 1.1.11. Denote $x = F_w(z)$ and $\alpha = (\alpha_1)^m$. For any $\tau \in \Lambda^n_{\alpha s,x}$, there exist $w^0, w^1, \ldots, w^k \in \Lambda^n_{\alpha s,x}$ such that $k \leq n$, $w^k = \tau$, $w^j \in \Lambda^j_{\alpha s,x}$ for $j = 0,1,\ldots,k$, where $\Lambda^0_{\alpha s,x} = \Lambda_{\alpha s,x}$, and $K_{w^{j-1}} \cap K_{w^j} \neq \emptyset$ for any $j = 1,\ldots,k$. Let $p = \max\{j | w^j \leq w\}$. Then, $w^j = wv^j$ for $j = 0,1,\ldots,p$. Since $|v^j| \geq m$, there exists $u^j \in \Lambda_\beta$ such that $v^j \leq u^j$. It follows that $z \in K_{v^0} \subseteq K_{u^0}$ and that $K_{u^{j-1}} \cap K_{u^j} \neq \emptyset$ for any $j = 1,\ldots,p$. Therefore, $u^j \in \Lambda^j_{\beta,z}$ for $j = 1,\ldots,p$. This implies that $K_{w^j} \subseteq F_w(U^j_\beta(z))$ for $j = 1,\ldots,p$. Hence $K_{w^p} \cap F_w(\overline{V}_0) = \emptyset$. If $p < k$, then there exists $w' \in \Lambda_s$ such that $w' \neq w$ and $w^{p+1} \leq w'$. Since $K_w \cap K_{w'} = F_w(V_0) \cap F_{w'}(V_0)$, we have $K_{w^p} \cap F_w(V_0) \neq \emptyset$. This is a contradiction and hence we have $p = k$. Hence $K_\tau \subseteq K_w$. Therefore, $U^{(n)}_{\alpha s}(x) \subseteq K_w$. □

PROOF OF THEOREM 1.3.11. (LF) $\wedge$ (GE) $\wedge$ $(\text{VD})_0 \Rightarrow (\text{A})_\text{n} \wedge (\text{VD})_0$: This is obvious by Theorem 1.3.8.
$(\text{A})_\text{n} \wedge (\text{VD})_0 \Rightarrow (\text{VD})_\text{n}$: For $s \in (0,1]$ and $x \in K$,

$$\mu(U^{(n)}_s(x)) \leq c_A\mu(K_s(x)) \leq c_Ac_B\mu(K_{\alpha s}(x)) \leq c_Ac_B\mu(U^{(n)}_{\alpha s}(x)).$$

$(\text{VD})_\text{n} \Rightarrow$ (GE): By Lemma 1.3.12, there exist $\alpha \in (0,1)$ and $z \in K$ such that $K_w \supseteq U^{(n)}_{\alpha s}(F_w(z))$ for any $s \in (0,1]$ and any $w \in \Lambda_s$. Proposition 1.3.7 implies that $\mu(U^{(n)}_s(x)) \leq c\mu(U^{(n)}_{\alpha s}(x))$. Now assume that $w \neq v \in \Lambda_s$ and $K_w \cap K_v \neq \emptyset$. Set $x = F_w(z)$. Then

$$\mu(K_v) \leq \mu(U^{(n)}_s(x)) \leq c\mu(U^{(n)}_{\alpha s}(x)) \leq c\mu(K_w). \tag{1.3.1}$$

Hence, μ is gentle.
$(\text{VD})_\text{n} \Rightarrow (\text{VD})_0$: Fix $\beta \in (0,1)$. Let $w \in \Lambda_s$. By Lemma 1.3.12, $K_{wv} \supset U^{(n)}_{\alpha\beta s}(x)$ for any $wv \in \Lambda_{\beta s}$, where $x = F_{wv}(z)$. Note that $K_w \subset U^{(n)}_s(x)$. By Proposition 1.3.7, there exists $c > 0$ such that $\mu(U^{(n)}_s(x)) \leq c\mu(U^{(n)}_{\alpha\beta s}(x))$. Hence $c\mu(K_{wv}) \geq c\mu(U^{(n)}_{\alpha\beta s}(x) \geq \mu(U^{(n)}_s(x)) \geq \mu(K_w)$. By Theorem 1.3.10, we have $(\text{VD})_0$.
$(\text{VD})_\text{n} \Rightarrow$ (LF): Let $s \in (0,1]$ and let $w \in \Lambda_s$. Choose $x = F_w(z)$, where z is given in Lemma 1.3.12. Then by (1.3.1), using the similar argument as in the proof of Theorem 1.3.8, we see that $\mu(K_v) \geq c^{-n}\mu(K_w)$ for any $v \in \Lambda^n_{s,x}$. Hence

$$c\mu(K_w) \geq c\mu(U^{(n)}_{\alpha s}(x)) \geq \mu(U^{(n)}_s(x)) = \sum_{v\in\Lambda^n_{s,x}} \mu(K_v) \geq c^{-n}\#(\Lambda^n_{s,x})\mu(K_w).$$

Dividing this by $\mu(K_w)$, we obtain (LF) by Lemma 1.3.6. □

Finally combining Theorems 1.3.10 and 1.3.11, we immediately obtain Theorem 1.3.5.

1.4. Locally finiteness and gentleness

In this section, we will define the notion of a scale being gentle with respect to another scale. It will turn out that the relation of "being gentle with respect to" is an equivalence relation among elliptic scales and the locally finiteness is inherited from a scale to another scale by this equivalence relation. As in the previous section, we fix a self-similar structure $\mathcal{L} = (K, S, \{F_i\}_{i \in S})$ with $K \neq \overline{V_0}$. Also all the scales are assumed to be right continuous.

DEFINITION 1.4.1. Let $\mathcal{S}_1$ and $\mathcal{S}_2$ be scales on Σ. $\mathcal{S}_2$ is said to be gentle with respect to $\mathcal{S}_1$ if and only if the gauge function of $\mathcal{S}_2$ is gentle with respect to $(\mathcal{L}, \mathcal{S}_1)$.

REMARK. Note that we need information on the self-similar structure $\mathcal{L}$ to determine whether $\mathcal{S}_2$ is gentle with respect to $\mathcal{S}_1$ or not.

Naturally we have the next proposition.

PROPOSITION 1.4.2. *Let $\mathcal{S}$ be a scale and let $\mu \in \mathcal{M}(K)$. Then μ is gentle with respect to $\mathcal{S}$ if and only if $\{\Lambda_s(\mu)\}_{0<s\leq 1}$ is gentle with respect to $\mathcal{S}$.*

Here is the main results of this section.

THEOREM 1.4.3. *Let $\mathcal{S}_1$ and $\mathcal{S}_2$ be elliptic scales on Σ. Assume that $\mathcal{S}_2$ is gentle with respect to $\mathcal{S}_1$.*
(1) $\mathcal{S}_1$ is gentle with respect to $\mathcal{S}_1$.
(2) If $\mathcal{S}_1$ is locally finite, then $\mathcal{S}_2$ is locally finite.
(3) $\mathcal{S}_1$ is gentle with respect to $\mathcal{S}_2$.
(4) Let $\mathcal{S}_3$ be an elliptic scale on Σ. Suppose $\mathcal{S}_3$ is gentle with respect to $\mathcal{S}_2$, then $\mathcal{S}_3$ is gentle with respect to $\mathcal{S}_1$.

PROOF. Let $\mathcal{S}_1 = \{\Lambda_s\}_{0<s\leq 1}$ and let $\mathcal{S}_2 = \{\Gamma_s\}_{0<s\leq 1}$. Let l and g be the gauge functions of $\mathcal{S}_1$ and $\mathcal{S}_2$ respectively. First we show (1). Recall that $w = w_1 \dots w_m \in \Lambda_s$ if and only if $l(w) \leq s < l(w_1 \dots w_{m-1})$. By Proposition 1.1.12-(1), there exists $c > 0$ such that $l(w) \geq cl(w_1 \dots w_{m-1})$ for any $w \in W_*$. Hence if $w \in \Lambda_s$, then $l(w) \leq s < l(w)/c$. This shows that $cl(w) \leq l(v)$ for any $w, v \in \Lambda_s$. Hence we obtain (1).

Proofs of (2), (3) and (4) are based on the same idea. If $w = w_1 \dots w_m \in \Gamma_s$, then

$$g(w_1 \dots w_{m-1}) > s \geq g(w).$$

On the other hand, there exists a unique $k \leq m$ such that $l(w_1 \dots w_{k-1}) > l(w_1 \dots w_k) = l(w_1 \dots w_m)$. By (EL2), $m - k \leq n$, where $n \in \mathbb{N}$ is independent of w. Let $a = l(w)$ and let $w' = w_1 \dots w_k$. Since $\mathcal{S}_2$ is gentle with respect to $\mathcal{S}_1$, $g(v) \leq cg(w')$ for any $v \in \Lambda_{a,w'}$. If $\mathcal{S}_1$ is elliptic, $g(w') \leq (\beta_1)^{-n} g(w)$, where β_1 is the constant appearing in Proposition 1.1.12-(1). Therefore there exists $c' > 0$ such that $g(v) \leq c'g(w)$ for any $v \in \Lambda_{a,w'}$. Using Proposition 1.1.12-(2), we see that there exists $p \in \mathbb{N}$ such that $g(v\tau) \leq g(w) \leq s$ for any $\tau \in W_p$. (Note that p in independent of s and w.) This shows that, for any $\tau \in W_p$ and any $v \in \Lambda_{a,w'}$, there exists a unique v' such that $v\tau \leq v'$ and $v' \in \Gamma_s$. Define $\pi(v\tau) = v'$. Then $\pi : \Lambda_{a,w'} \times W_p \to \Gamma_s$. Note that $\Gamma_{s,w}$ is included in the image of π. Hence $\#(\Gamma_{s,w}) \leq \#(\Lambda_{a,w'})N^p$, where $N = \#(S)$. By Lemma 1.3.6, if $\mathcal{S}_1$ is locally finite, then so is $\mathcal{S}_2$. This proves (2). Next we show (3). For any $v' \in \Gamma_{s,w}$, choose $v \in \Lambda_{a,w'}$ and $\tau \in W_p$ so that $\pi(v\tau) = v'$. Then $l(v\tau) \leq l(v')$. Since $\mathcal{S}_1$ satisfies (EL1), there exists $\gamma > 0$ such that $l(wi) \geq \gamma l(w)$ for any $w \in W_*$ and any $i \in S$. Therefore,

$$l(v') \geq l(v\tau) \geq \gamma^p l(v) \geq \gamma^p a = \gamma^p l(w).$$

Hence $\mathcal{S}_1$ is gentle with respect to $\mathcal{S}_2$.

To prove (4), we write $\mathcal{S}_3 = \{\Omega_s\}_{0<s\leq 1}$. Let $w \in \Omega_s$. There exist $k, j \in \mathbb{N}$ such that $j \leq k \leq m = |w|$, $g(w_1 \dots w_{k-1}) > g(w_1 \dots w_k) = g(w)$ and $l(w_1 \dots w_{j-1}) >$

$l(w_1 \dots w_j) = l(w_1 \dots w_k)$. Now using the same construction as π above, we have maps $\pi_1 : \Lambda_{a,w''} \times W_p \to \Gamma_b$ and $\pi_2 : \Gamma_{b,w'} \times W_q \to \Omega_s$, where $w'' = w_1 \dots w_j, a = l(w''), w' = w_1 \dots w_k$ and $b = g(w')$, satisfying the same properties as π. Now for any $v \in \Omega_{s,w}$, there exist $v' \in \Gamma_{b,w'}$, $\tau' \in W_q$, $v'' \in \Lambda_{a,w''}$ and $\tau'' \in W_p$ such that $\pi_2(v'\tau') = v$ and $\pi_1(v''\tau'') = v'$. Note that $v'\tau' \le v$ and $v''\tau'' \le v'$. This implies

$$l(v) \ge l(v'\tau') \ge \gamma^q l(v') \ge \gamma^q l(v''\tau'') \ge \gamma^{p+q} l(v'') \ge \gamma^{p+q} a \ge \gamma^{p+q} l(w)$$

This shows that $\mathcal{S}_1$ is gentle with respect to $\mathcal{S}_3$. Applying (3), we obtain the desired result. □

By the above theorem, the relation "gentle with respect to" is an equivalence relation on elliptic scales.

DEFINITION 1.4.4. (1) Let $\mathcal{S}_1$ and $\mathcal{S}_2$ be elliptic scales. We write $\mathcal{S}_1 \underset{\text{GE}}{\sim} \mathcal{S}_2$ if and only if $\mathcal{S}_1$ is gentle with respect to $\mathcal{S}_2$.
(2) Let $\mathcal{S}$ be a scale. We define

$$\mathcal{M}_{\text{VD}}(\mathcal{L}, \mathcal{S}) = \{\mu | \mu \in \mathcal{M}(K), \mu \text{ has (VD) with respect to } \mathcal{S}\}.$$

PROPOSITION 1.4.5. (1) *Let $\mathcal{S}$ be an elliptic scale on Σ. If $\mathcal{M}_{\text{VD}}(\mathcal{L}, \mathcal{S}) \neq \emptyset$, then $\mathcal{S}$ is locally finite.*
(2) *Let $\mathcal{S}_1$ and $\mathcal{S}_2$ be elliptic scales. Then*

$$\mathcal{M}_{\text{VD}}(\mathcal{L}, \mathcal{S}_1) \cap \mathcal{M}_{\text{VD}}(\mathcal{L}, \mathcal{S}_2) \neq \emptyset \Rightarrow \mathcal{S}_1 \underset{\text{GE}}{\sim} \mathcal{S}_2 \Rightarrow \mathcal{M}_{\text{VD}}(\mathcal{L}, S_1) = \mathcal{M}_{\text{VD}}(\mathcal{L}, S_2).$$

PROOF. (1) This is immediate by Theorem 1.3.5.
(2) Let $\mu \in \mathcal{M}_{\text{VD}}(\mathcal{L}, \mathcal{S}_1) \cap \mathcal{M}_{\text{VD}}(\mathcal{L}, \mathcal{S}_2)$. If $\mathcal{S}_3$ is the scale induced by μ, then $\mathcal{S}_1 \underset{\text{GE}}{\sim} \mathcal{S}_3$ and $\mathcal{S}_2 \underset{\text{GE}}{\sim} \mathcal{S}_3$ by Proposition 1.4.2. Hence $\mathcal{S}_1 \underset{\text{GE}}{\sim} \mathcal{S}_2$. Next assume $\mathcal{S}_1 \underset{\text{GE}}{\sim} \mathcal{S}_2$ and let $\mu \in \mathcal{M}_{\text{VD}}(\mathcal{L}, \mathcal{S}_1)$. Let $\mathcal{S}_3$ be the scale induced by μ. Then $\mathcal{S}_1 \underset{\text{GE}}{\sim} \mathcal{S}_3$ by Proposition 1.4.2. Hence $\mathcal{S}_2 \underset{\text{GE}}{\sim} \mathcal{S}_3$. Again by Proposition 1.4.2, $\mu \in \mathcal{M}_{\text{VD}}(\mathcal{L}, \mathcal{S}_2)$. Hence $\mathcal{M}_{\text{VD}}(\mathcal{L}, \mathcal{S}_1) \subseteq \mathcal{M}_{\text{VD}}(\mathcal{L}, \mathcal{S}_2)$. Exchanging $\mathcal{S}_1$ and $\mathcal{S}_2$, we see $\mathcal{M}_{\text{VD}}(\mathcal{L}, \mathcal{S}_1) = \mathcal{M}_{\text{VD}}(\mathcal{L}, \mathcal{S}_2)$. □

Denote the collection of elliptic scales on Σ by $\mathcal{ES}(\Sigma)$. Then, by the above results, an equivalence class of $\mathcal{ES}(\Sigma)/\underset{\text{GE}}{\sim}$ tells us whether a scale $\mathcal{S}$ is locally finite or not and determines $\mathcal{M}_{\text{VD}}(\mathcal{L}, \mathcal{S})$, the family of volume doubling measures with respect to $\mathcal{S}$. Those facts raises our curiosity on the structure of $\mathcal{ES}(\Sigma)/\underset{\text{GE}}{\sim}$. In the following sections, we will study this problem in a restricted situation.

We conclude this section by giving an important necessary condition for two self-similar scales being gentle.

NOTATION. For $w \in W_\#$ and any $n \in \mathbb{N}$, we define $(w)^n = \underbrace{w \dots w}_{n \text{ times}} \in W_*$. Also $(w)^\infty = www \dots \in \Sigma$.

LEMMA 1.4.6. *Let $\mathbf{a} = (a_i)_{i \in S} \in (0,1)^S$ and let $\mathbf{b} = (b_i)_{i \in S} \in (0,1)^S$. Assume that $\mathcal{S}(\mathbf{a}) \underset{\text{GE}}{\sim} \mathcal{S}(\mathbf{b})$. If $w, w', v, v' \in W_\#$ and $\pi(v(w)^\infty) = \pi(v'(w')^\infty)$, then $\dfrac{\log a_w}{\log b_w} = \dfrac{\log a_{w'}}{\log b_{w'}}$.*

PROOF. For sufficiently small s, there exist $i(s), j(s) \in \mathbb{N}$ and $w(s), w'(s) \in W_*$ such that $w < w(s), w' < w'(s)$, $v(w)^{i(s)}w(s), v'(w')^{j(s)}w'(s) \in \Lambda_s(\mathbf{a})$ Set $v(s) = v(w)^{i(s)}w(s)$ and $v'(s) = v'(w')^{j(s)}w'(s)$. By Lemma 1.1.13, $a_{v(s)}/a_{v'(s)}$ is uniformly bounded with respect to s. Since $a_v a_{w(s)}/(a_{v'}a_{w'(s)})$ is uniformly bounded, we see that $(a_w)^{i(s)}/(a_{w'})^{j(s)}$ is uniformly bounded with respect to s. As $\mathcal{S}(\mathbf{a}) \underset{\mathrm{GE}}{\sim} \mathcal{S}(\mathbf{b})$, we see that $b_{v(s)}/b_{v'(s)}$ is uniformly bounded as well. Note that $b_v b_{w(s)}/(b_{v'}b_{w'(s)})$ is uniformly bounded. Hence

$$\frac{(b_w)^{i(s)}}{(b_{w'})^{j(s)}} = (a_w)^{i(s)(\alpha-\beta)}\frac{(a_w)^{i(s)\beta}}{(a_{w'})^{j(s)\beta}},$$

where $\alpha = \log b_w/\log a_w$ and $\beta = \log b_{w'}/\log a_{w'}$, is uniformly bounded. Since $i(s) \to +\infty$ as $s \downarrow 0$, it follows that $\alpha = \beta$. □

1.5. Rationally ramified self-similar sets 1

In this section, we will introduce a special class of self-similar structures called "rationally ramified self-similar structures". $\mathcal{L} = (K, S, \{F_i\}_{i \in S})$ is assumed to be a self-similar structure throughout this section. Roughly speaking, $\mathcal{L}$ is called rationally ramified if $K_i \cap K_j$ is again a self-similar set. This class of self-similar sets include post critically finite self-similar sets, for example, the Sierpinski gasket, as well as so called "infinitely ramified" self-similar sets like the Sierpinski carpet and the Menger sponge. The advantage of a rationally ramified self-similar structure is that one can give simple characterizations for the locally finiteness of a scale and the gentleness of two scales. Using such results, we can explicitly determine the class of self-similar measures which have the volume doubling property with respect to a given scale for rationally ramified self-similar sets. See the next section for details

To start with, we need several notions and results on the shift space.

DEFINITION 1.5.1. Let X be a non-empty finite subset of $W_\#$. For $w \in W_\#$, we denote $w = (w)_1 \dots (w)_{|w|}$, where $(w)_i \in S$ for $i = 1, \dots, |w|$. We define a map ι_X from $\Sigma(X) = \{x_1x_2\dots | x_i \in X \text{ for any } i \in \mathbb{N}\}$ to $\Sigma(S)$ by

$$\iota_X(x_1x_2\dots) = (x_1)_1 \dots (x_1)_{|x_1|}(x_2)_1 \dots (x_2)_{|x_2|} \cdots.$$

Define $\Sigma[X] = \iota_X(\Sigma(X))$, $K[X] = \pi(\Sigma[X])$, $\Sigma_w[X] = \sigma_w(\iota_X(\Sigma(X)))$ and $K_w[X] = F_w(K[X]) (= \pi(\Sigma_w[X]))$ for $w \in W_\#$. X is called independent if and only if ι_X is injective. When X is independent, we sometimes identify $\Sigma(X)$ with $\Sigma[X]$.

For example, let $S = \{1, 2\}$ and let $X = \{1, 12, 21\}$ Then X is not independent. In fact,

$$2112(1)^\infty = \iota_X(cb(a)^\infty) = \iota_X(cac(a)^\infty).$$

where $a = 1, b = 12, c = 21$.

Since $\iota_X : \Sigma(X) \to \Sigma[X]$ and $\pi : \Sigma \to K$ are continuous, we have the following lemma.

LEMMA 1.5.2. *If X is a nonempty finite subset of W_*, then $\Sigma[X]$ and $K[X]$ are compact.*

Now we study how to characterize the independence of X.

DEFINITION 1.5.3. Let X be a nonempty subset of $W_\#$.
(1) For $m \geq 0$, define $\rho_m : \Sigma \to W_m$ by $\rho_m(\omega) = \omega_1 \dots \omega_m$ for $\omega = \omega_1\omega_2 \dots$.

(2) For $m \geq n$, define $\rho_{m,n} : W_m \to W_n$ by $\rho_{m,n}(w_1 \dots w_m) = w_1 \dots w_n$.
(3) For $x_1, x_2, \dots, x_m \in X$, recalling that each $x_i \in W_*$, we may regard $x_1 \dots x_m \in W_*(X)$ as an element of W_*. We use $\iota_X^w(x_1 \dots x_m)$ to denote $x_1 \dots x_m$ as an element of W_* to avoid confusion. In other word, $\iota_X^w : W_*(X) \to W_*$ is defined by

$$\iota_X^w(x_1 \dots x_m) = x_1(1) \dots x_1(n_1) \dots x_m(1) \dots x_m(n_m),$$

where $n_i = |x_i|$ and $x_i = x_i(1)x_i(2) \dots x_i(n_i) \in W_{n_i}$ for $i = 1, \dots, m$.
(4) For $m \geq 0$, we define

$$Q_m(X) = \bigcup_{w,v \in X, w \neq v} \Big(\rho_m(\Sigma_w[X])) \cap \rho_m(\Sigma_v[X]) \Big).$$

The following fact is immediate from the above definition. It says that an element of $Q_m(X)$ can be expressed by two different words of X whose first symbols are different. X is assumed to be a nonempty finite subset of $W_\#$.

LEMMA 1.5.4. *Let $w \in W_m$. Then $w \in Q_m(X)$ if and only if there exist $x_1 \dots x_k$ and $x'_1 \dots x'_n \in W_*(X)$ such that $x_1 \neq x'_1, \iota_X^w(x_1 \dots x_k) \leq w < \iota_X^w(x_1 \dots x_{k-1})$ and $\iota_X^w(x'_1 \dots x'_n) \leq w < \iota_X^w(x'_1 \dots x'_{n-1})$. Moreover, if $m \geq n$, then $\rho_{m,n}(Q_m(X)) \subseteq Q_n(X)$.*

LEMMA 1.5.5. *Let $\omega \in \Sigma$. Suppose that there exist $w \in X$ and $m_1 < m_2 < \dots$ such that $\rho_{m_i}(\omega) \in \rho_{m_i}(\Sigma_w[X])$. Then $\omega = \iota_X(wx_2 \dots)$ for some $x_2, x_3, \dots \in X$.*

PROOF. For sufficiently large i, we may find $x(i) \in X$ such that $\rho_{m_i}(\omega) \in \Sigma_{wx(i)}$. Since X is a finite set, we may find $x_2 \in X$ and a subsequence $\{m_{2,i}\}_{i \geq 1}$ of $\{m_i\}_{i \geq 1}$ such that $\rho_{m_{2,i}}(\omega) \in \Sigma_{wx_2}$ for any i. Repeating the same procedure, we may inductively obtain $x_j \in X$ and $\{m_{j,i}\}_{i \geq 1}$ for $j \geq 2$. Now, $\omega = \iota_X(wx_2x_3 \dots)$. □

We have a simple characterization of the independence of X in terms of $Q_m(X)$.

THEOREM 1.5.6. *Let X be a nonempty finite subset of $W_\#$. Then X is independent if and only if $Q_m(X) = \emptyset$ for some $m \in \mathbb{N}$.*

REMARK. By Lemma 1.5.4, if $Q_m(X) = \emptyset$, then $Q_n(X) = \emptyset$ for any $n \geq m$.

PROOF. If X is not independent, then there exist $x_1x_2 \dots, x'_1x'_2 \dots \in \Sigma(X)$ such that $\iota_X(x_1x_2 \dots) = \iota_X(x'_1x'_2 \dots)$. We may assume that $x_1 \neq x'_1$ without loss of generality. Now, $\rho_m(\iota_X(x_1x_2 \dots)) \in Q_m(X)$ for any $m \geq 0$.

Conversely suppose that $Q_m(X) \neq \emptyset$ for any $m \geq 0$. Set

$$Q^*_{m,n}(X) = \rho_{m+n,m}(Q_{m+n}(X)).$$

Then $\{Q^*_{m,n}(X)\}_{n \geq 0}$ is a decreasing sequence of nonempty finite sets. Therefore, $Q^*_m(X) = \cap_{n \geq 0} Q^*_{m,n}(X)$ is not empty. Also it follows that $\rho_{k,l}(Q^*_k(X)) = Q^*_l(X)$ for any $k, l \in \mathbb{N}$ with $k \geq l$. Therefore, there exists $\omega = \omega_1\omega_2 \dots \in \Sigma$ such that $\rho_m(\omega) \in Q^*_m(X)$ for any $m \geq 0$. For each m, there exist $w(m), v(m) \in X$ such that $w(m) \neq v(m)$ and $\rho_m(\omega) \in \rho_m(\iota_X(\Sigma(X))) \cap \Sigma_{w(m)} \cap \Sigma_{v(m)}$. Since X is a finite set, there exist $w, v \in X$ and $\{m_i\}_{i \geq 1}$ such that $w \neq v$, $w(m_i) = w$ and $v(m_i) = v$. Now using Lemma 1.5.5, we see that $\omega = \iota_X(wx_2x_3 \dots) = \iota_X(vx'_2x'_3 \dots)$ for some $\{x_i\}, \{x'_i\} \in X$. Hence ι_X is not injective and therefore X is not independent. □

Hereafter, if a nonempty finite subset X of $W_\#$ is independent, we think of $x_1 \dots x_m \in W_*(X)$ (where $x_i \in X$ for any i) as an element of W_* in the natural manner.

Before getting to the definition of rationally ramified self-similar structure, we still need several notions.

DEFINITION 1.5.7. Let Σ_0 be a nonempty subset of Σ and let $x \in W_*$. We define $O_{\Sigma_0,x}(\omega)$

$$O_{\Sigma_0,x}(\omega) = \#(\{m | m \geq 0, \sigma^m \omega \in \sigma_x(\Sigma_0)\})$$

for any $\omega \in \Sigma$. We allow ∞ as a value of $O_{\Sigma_0,x}(\omega)$.

The following two lemmas are basic facts on $O_{\Sigma[X],x}(\omega)$.

LEMMA 1.5.8. *Let X be a nonempty finite subset of W_* and let $x \in W_*$. If*

$$\sup_{\omega\in\Sigma} O_{\Sigma[X],x}(\omega) = +\infty,$$

then there exists $\tau \in \Sigma$ such that $O_{\Sigma[X],x}(\tau) = \infty$.

PROOF. Set $k = \max_{w\in X} |w|$ and define $M = |x|(k^2 + 3)$. By the above assumption, we may choose $\omega \in \Sigma_x[X]$ so that $\#(\{m|\sigma^m(\omega) \in \Sigma_x[X]\}) \geq M$. Let $\omega = x\iota_X(x_1 x_2 \dots)$, where $x_1, x_2, \dots \in X$. There exists a sequence $\{m_i\}_{0=1,\dots,k^2+2}$ such that $m_0 = 0$, $m_i + |x| \leq m_{i+1}$ for any $i = 1, \dots, k^2 + 1$ and $\sigma^{m_i}(\omega) \in \Sigma_x[X]$ for any $i = 1, \dots, k^2 + 2$. Choose n so that $m_{k^2+2} + |x| < |xx_1 \dots x_{n-1}|$. Then $|x_{n+1} \dots x_{n+k}| \geq k$ and so, for any $i = 1, \dots, k^2 + 2$, there exists $\{x_j^i\}_{j=1,\dots,n(i)} \subset X$ such that $xx_1 \dots x_n x_{n+1} \dots x_{n+k} < \omega_1 \dots \omega_{m_i} x x_1^i \dots x_{n(i)}^i \leq xx_1 \dots x_n$. Since $k^2 + 2 > |x_{n+1} \dots x_{n+k}| + 1$, we see $\omega_1 \dots \omega_{m_p} x x_1^p \dots x_{n(p)}^p = \omega_1 \dots \omega_{m_q} x x_1^q \dots x_{n(q)}^q$ for some $0 \leq p < q \leq k$. (We set $x_i^0 = x_i$ and $n(0) = n$.) Note that $m_p + |x| \leq m_q$. Hence we have l which satisfies $\sigma^l(x_1^p \dots x_{n(p)}^p) = xx_1^q \dots x_{n(q)}^q$. Set $w = x_1^p \dots x_{n(p)}{}^p$ and define $\tau = (w)^\infty$. Then $\tau = (w)^i w_1 \dots w_l x x_1^q \dots x_{n(q)}^q (w)^\infty$ for any i. Therefore, $\sigma^{|w|i+l}\tau \in \Sigma_x[X]$ for any i. □

REMARK. In the proof, we have shown the following statement:
Let $k = \max_{w\in X} |w|$. If $O_{\Sigma[X],x}(\omega) \geq M = |x|(k^2 + 3)$ for some $\omega \in \Sigma$, then there exists $\tau \in \Sigma$ such that $O_{\Sigma[X],x}(\tau) = +\infty$.

As a final step to the definition of rationally ramified self-similar structures, we need several definitions.

DEFINITION 1.5.9. Let $\Omega = (X, Y, \varphi, x, y)$, where X and Y are a non-empty independent finite subsets of $W_\#$, φ is a bijective map between X and Y and $x, y \in W_\#$.
(1) We define $\varphi_* : \Sigma_x[X] \to \Sigma_y[Y]$ by $\varphi_*(xx_1x_2 \dots) = y\varphi(x_1)\varphi(x_2) \dots$ for any $x_1, x_2, \dots \in X$.
(2) A pair $(\omega, \tau) \in \Sigma(S) \times \Sigma(S)$ with $\omega \neq \tau$ is called a corresponding pair with respect to Ω if and only if $\omega = v\omega'$ and $\tau = v\varphi_*(\omega')$ for some $v \in W_*$ and some $\omega' \in \Sigma_x[X]$.
(3) Ω is called a relation of $\mathcal{L}$ if and only if the first symbol of x is different from that of y, $O_{\Sigma[X],x}(\omega)$ and $O_{\Sigma[Y],y}(\omega)$ are finite for any $\omega \in \Sigma$ and $\pi(\omega) = \pi(\tau)$ for any corresponding pair (ω, τ) with respect to Ω. The collection of relations of $\mathcal{L}$ is denoted by $\mathcal{R}_{\mathcal{L}}$.

(4) Let $\Omega = (X, Y, \varphi, x, y)$ be a relation of $\mathcal{L}$. $\Omega' = (X', Y', \varphi', x, y)$ is called a sub-relation of Ω if $X' \subseteq X$, $Y' = \varphi(X')$ and $\varphi' = \varphi|_X$.
(5) Let $\mathcal{R} \subset \mathcal{R}_{\mathcal{L}}$. A relation $\Omega = (X, Y, \varphi, x, y)$ is said to be generated by $\mathcal{R}$ if there exists a sequence of sub-relations of relations in $\mathcal{R}$, $\{(X_i, X_{i+1}, \varphi_i, x_i, x_{i+1})\}_{i=1}^{m-1}$, such that $X = X_1, Y = X_m, x = x_1, y = x_m$ and $\varphi = \varphi_{m-1} \circ \ldots \circ \varphi_1$. We use $[\mathcal{R}]$ to denote the collection of relations generated by $\mathcal{R}$. If $\mathcal{R} \subseteq \mathcal{R}' \subseteq [\mathcal{R}]$, then $\mathcal{R}'$ is said to be generated by $\mathcal{R}$ or $\mathcal{R}$ is a generator of $\mathcal{R}'$.

REMARK. If (X, Y, φ, x, y) is a relation of $\mathcal{L}$. Then $\sup_{w \in \Sigma} O_{\Sigma[X],x}(w)$ and $\sup_{w \in \Sigma} O_{\Sigma[Y],y}(w)$ is finite by Lemma 1.5.8.

REMARK. If $\Omega = (X, Y, \varphi, x, y)$ be a relation of $\mathcal{L}$, then so is $(Y, X, \varphi^{-1}, y, x)$. We denote $\Omega^{-1} = (Y, X, \varphi^{-1}, y, x)$ and identify Ω with Ω^{-1}. In particular, if $\mathcal{R}$ is a subset of $\mathcal{R}_{\mathcal{L}}$ for a self-similar structure $\mathcal{L}$, then we always suppose that $\Omega^{-1} \in \mathcal{R}$ for any $\Omega \in \mathcal{R}$. In making a list of elements of a relation set, we customary mention only one of (X, Y, φ, x, y) or $(Y, X, \varphi^{-1}, y, x)$.

DEFINITION 1.5.10 (Rationally ramified self-similar structure). A self-similar structure $\mathcal{L} = (K, S, \{F_i\}_{i \in S})$ is said to be rationally ramified if and only if it is strongly finite and there exists a finite subset $\mathcal{R}$ of $\mathcal{R}_{\mathcal{L}}$ satisfying the following property: for any $i, j \in S$ with $i \neq j$,

$$\pi^{-1}(K_i \cap K_j) \cap \Sigma_i = \bigcup_{(X,Y,\varphi,x,y) \in \mathcal{R}_{ij}, x \in \sigma_i(W_*)} \Sigma_x[X], \tag{1.5.1}$$

where

$$\mathcal{R}_{ij} = \{\Omega | \Omega = (X, Y, \varphi, x, y) \in [\mathcal{R}], x \in \sigma_i(W_*), y \in \sigma_j(W_*)\}.$$

$\mathcal{R}$ is called the relation set of $\mathcal{L}$.

Note that $[\mathcal{R}]$ is a finite set if $\mathcal{R}$ is finite. We may assume that $\mathcal{R} = [\mathcal{R}]$ in the above definition without loss of generality. However, as one will see in Example 1.5.12, $\mathcal{R}$ can be more simple than $[\mathcal{R}]$ in some cases.

EXAMPLE 1.5.11 (the Sierpinski gasket). Let p_1, p_2 and p_3 be vertices of a regular triangle in $\mathbb{C}$. Define $F_i(z) = (z - p_i)/2 + p_i$ for $i = 1, 2, 3$. The Sierpinski gasket is the self-similar set with respect to $\{F_1, F_2, F_3\}$, i.e. K is the unique non-empty compact set satisfying $K = F_1(K) \cup F_2(K) \cup F_3(K)$. $\mathcal{L} = (K, S, \{F_i\}_{i \in S})$, where $S = \{1, 2, 3\}$, is a rationally ramified self-similar structure. Indeed, $\{(\{i\}, \{j\}, \varphi_{ij}, j, i) | (i, j) = (1, 2), (2, 3), (3, 1)\}$, where $\varphi_{ij}(i) = j$, is a relation set. According to the convention in the remark above, this relation set contains 6 elements in fact.

EXAMPLE 1.5.12 (the Sierpinski carpet). et $p_1 = 0, p_2 = 1/2, p_3 = 1, p_4 = 1 + \sqrt{-1}/2, p_5 = 1 + \sqrt{-1}, p_6 = 1/2 + \sqrt{-1}, p_7 = \sqrt{-1}$ and $p_8 = \sqrt{-1}/2$. Define $F_i : \mathbb{C} \to \mathbb{C}$ by $F_i(z) = (z - p_i)/3 + p_i$ for $i = 1, \ldots, 8$. Then there exists a unique nonempty compact subset to $\mathbb{C}$, K, which satisfies $K = \cup_{i=1}^{8} F_i(K)$. K is called the Sierpinski carpet. Let $\mathcal{L} = (K, S, \{F_i\}_{i \in S})$, where $S = \{1, \ldots, 8\}$. Then $\mathcal{L}$ is a rationally ramified self-similar structure. To describe its relation set $\mathcal{R}$, we let $X_1 = \{1, 2, 3\}, Y_1 = \{7, 6, 5\}, \varphi_1(1) = 7, \varphi_1(2) = 6, \varphi_1(3) = 5, X_2 = \{1, 8, 7\}, Y_2 = \{3, 4, 5\}, \varphi_2(1) = 3, \varphi_2(8) = 4$ and $\varphi_2(7) = 5$. Then

$$\mathcal{R} = \{(X_1, Y_1, \varphi_1, i, j) | (i, j) = (8, 1), (4, 3), (7, 8), (5, 4)\} \cup \{(X_2, Y_2, \varphi_2, i, j) | (i, j) = (2, 1), (6, 7), (3, 2), (5, 6)\},$$

FIGURE 1.1. Sierpinski gasket

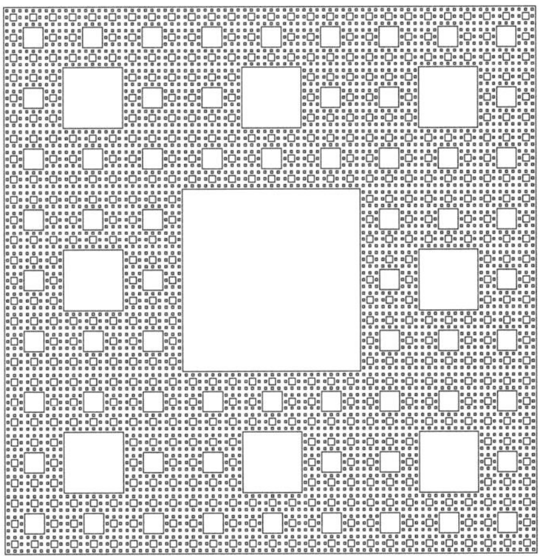

FIGURE 1.2. Sierpinski carpet

where $\varphi_3(1) = 5$ and $\varphi_4(3) = 7$. In this case, the set of relations generated by $\mathcal{R}$, $[\mathcal{R}]$, is not equal to $\mathcal{R}$. In fact,

$$[\mathcal{R}] = \mathcal{R} \cup \{(\{1\}, \{5\}, \varphi_{15}, i, j), (\{3\}, \{7\}, \varphi_{37}, k, l) | \\ (i, j) = (6, 8), (4, 2), (k, l) = (8, 2), (6, 4)\},$$

where φ_{mn} maps m to n. Those additions are really needed in the definition of rationally ramified self-similar structure. For example, $\mathcal{R}_{42} = \{(\{1\}, \{5\}, \varphi_{15}, 4, 2)\}$.

PROPOSITION 1.5.13. *Let $\mathcal{L} = (K, S, \{F_i\}_{i \in S})$ be a rationally ramified self-similar structure with a relation set $\mathcal{R}$.*
(1) $K \neq \overline{V}_0$.
(2) Set $M = \max_{x \in K} \#(\pi^{-1}(x))$. Suppose that $\pi(\omega) = \pi(\tau)$ and $\omega \neq \tau$. Then there exist $\Omega_1, \ldots, \Omega_m \in [\mathcal{R}]$ and $\omega^{(1)}, \ldots, \omega^{(m+1)} \in \Sigma$ which satisfy the following conditions (AS1), (AS2) and (AS3)*:*

(AS1) $m + 1 \leq \max_{x \in K} \#(\pi^{-1}(x))$
(AS2) $\omega = \omega^{(1)}, \omega^{(m+1)} = \tau$ *and* $(\omega^{(i)}, \omega^{(i+1)})$ *is a corresponding pair with respect to* Ω_i *for any* $i = 1, \ldots, m$.
(AS3) $s(\omega^{(i)}, \tau) < s(\omega^{(i+1)}, \tau)$ *for any* $i = 1, \ldots, m-2$, *where* $s(\delta, \rho) = \min\{k - 1 | \delta_k \neq \rho_k\}$ *for* $\delta = \delta_1\delta_2\ldots$ *and* $\rho = \rho_1\rho_2\ldots$.

REMARK. Under the assumptions of the above proposition, let $\omega^{(1)}, \ldots, \omega^{(m+1)}$ and $\Omega_1, \ldots, \Omega_m$ satisfy (AS1), (AS2) and (AS3). Set $m_n = s(\tau, \omega^{(n)})$. If $\Omega_n = (X_n, Y_n, \varphi_n, x(n), y(n))$, then the first symbols of $x(i)$ and $y(i)$ are $\omega^{(n)}_{m_n+1}$ and τ_{m_n+1} respectively. Furthermore, $\omega^{(n)} = \tau_1 \ldots \tau_{m_n} x(n) x_1 x_2 \ldots$ for some $x_1 x_2 \cdots \in \Sigma[X]$ and $\omega^{(n+1)} = \tau_1 \ldots \tau_{m_n} y(n) \varphi_n(x_1) \varphi_n(x_2) \ldots$.

PROOF. (1) Since $\mathcal{C}_{\mathcal{L}} = \cup_{(X,Y,\varphi,x,y) \in \mathcal{R}} \Sigma_x[X]$, the post critical set $\mathcal{P}_{\mathcal{L}}$ is a finite union of $\Sigma_w[X]$'s for some $w \in W_*$ and some X where $(X, Y, \varphi, x, y) \in \mathcal{R}$. Now V_0 is a finite union of $F_w(K[X])$'s. Lemma 1.5.2 shows that $V_0 = \overline{V}_0$. By [**28**, Corollaries 1.4.8 and 1.4.9], $\mathcal{L}$ is minimal. Hence, [**28**, Theorem 1.3.8] implies that $K \neq V_0 = \overline{V}_0$.
(2) Define $\omega^{(1)}, \omega^{(2)}, \ldots$ and $\Omega_1, \Omega_2, \ldots$ inductively as follows. Set $\omega^{(1)} = \omega$. Suppose we have $\omega^{(1)}, \ldots, \omega^{(n)}$ and $\Omega_1, \ldots, \Omega_{n-1}$. If $\omega^{(n)} = \tau$, then we set $m = n+1$ and finish the construction. If $\omega^{(n)} \neq \tau$, then set $k = s(\omega^{(n)}, \tau) + 1$, $i = \omega^{(n)}_k$ and $j = \tau_k$. By (1.5.1), we may choose $\Omega_n = (X, Y, \varphi, x, y) \in \mathcal{R}_{ij}$ such that $\sigma^k(\omega^{(n)}) \in \Sigma_x[X]$. Define $\omega^{(n+1)} = \tau_1 \ldots \tau_k \varphi_*(\sigma^k(\omega^{(n)}))$. Then $\omega^{(n+1)}_{k+1} = \tau_{k+1}$. Hence $s(\omega^{(n+1)}, \tau) > k$. As far as this construction continues, $\omega^{(1)}, \omega^{(2)}, \ldots, \omega^{(n)}$ and τ are mutually different elements. Therefore, $n+1$ will not exceed $\max_{x \in K} \#(\pi^{-1}(x))$. □

The next two lemmas describe fine structures of intersections of two copies K_w and K_v for a rationally ramified self-similar set. They are technically useful in getting results in the following sections.

LEMMA 1.5.14. *Let* $\mathcal{L}$ *be rationally ramified and let* $(X, Y, \varphi, x, y) \in \mathcal{R}_{\mathcal{L}}$. *Define* $\hat{\varphi} : \Sigma[X] \to \Sigma[Y]$ *by* $\hat{\varphi}(x_1 x_2 \ldots) = \varphi(x_1)\varphi(x_2)\ldots$. *Then there exists a unique homeomorphism* $\tilde{\varphi} : K[X] \to K[Y]$ *that satisfies* $\tilde{\varphi} \circ \pi = \pi \circ \hat{\varphi}$.

PROOF. Fix $p \in K[X]$. If there exist $x_1 x_2 \ldots \in \Sigma(X)$ and $x'_1 x'_2 \ldots \in \Sigma(X)$ such that $\pi(x_1 x_2 \ldots) = \pi(x'_1 x'_2 \ldots) = p$, then $\pi(y\hat{\varphi}(x_1 x_2 \ldots)) = \pi(x x_1 x_2 \ldots) = \pi(x x'_1 x'_2 \ldots) = \pi(y\hat{\varphi}(x'_1 x'_2 \ldots))$. Therefore, $\pi(\hat{\varphi}(x_1 x_2 \ldots)) = \pi(\hat{\varphi}(x'_1 x'_2 \ldots))$. Hence for any $p \in K[X]$, $\pi(\hat{\varphi}(\pi^{-1}(x)))$ contains only one point. Define $\tilde{\varphi}(p)$ as this one point. Then by a routine argument, $\tilde{\varphi} : K[X] \to K[Y]$ is continuous. Exchanging X and Y, we obtain the inverse of $\tilde{\varphi}$. Hence $\tilde{\varphi}$ is a homeomorphism. □

DEFINITION 1.5.15. Let X be a finite subset of $W_\#$ and let $x \in W_*$. For each $w \in W_*$, define

$$A_{X,x}(w) = \{(z, x_0, x_1, \ldots, x_m) | m \geq 0, z \in W_*,$$
$$x_0 = x, x_1, \ldots, x_m \in X, z x_0 x_1 \ldots x_m \leq w < z x_0 x_1 \ldots x_{m-1}\}.$$

LEMMA 1.5.16. *Let* $\mathcal{L}$ *be rationally ramified and let* $\Omega = (X, Y, \varphi, x, y) \in \mathcal{R}_{\mathcal{L}}$. *Suppose* $w = w_1 \ldots w_m, v = v_1 \ldots v_n \in W_\#$ *and* $\Sigma_w \cap \Sigma_v = \emptyset$. *Set* $z_* = w_1 \ldots w_N$, *where* $N = \inf\{i | w_i = v_i\} - 1$. *Then, there exist a corresponding pair with respect to* Ω *in* $\Sigma_w \times \Sigma_v$ *if and only if there exist* $(z_*, x, x_1, \ldots, x_m) \in A_{X,x}(w)$ *and*

$(z_*, y, y_1, \ldots, y_n) \in A_{Y,y}(v)$ such that $y_i = \varphi(x_i)$ for $i = 1, \ldots, \min(m, n)$. Moreover, let $x_m = x_m^1 x_m^2$ where $w = zxx_1 \ldots x_{m-1}x_m^1$ and let $y_n = y_n^1 y_n^2$ where $v = zyy_1 \ldots y_{n-1}y_n^1$. Suppose that $m \geq n$ and define $y_{n+1}, \ldots, y_m$ by $y_i = \varphi(x_i)$. Then $K_{wx_m^2}[X] = K_{vy_n^2 y_{n+1} \ldots y_m}[Y] \subseteq K_w \cap K_v$ and $(F_v)^{-1} \circ F_w|_{K_{x_m^2}[X]} = F_{y_n^2 y_{n+1} \ldots y_m} \circ \tilde{\varphi} \circ (F_{x_m^2})^{-1}$, where $\tilde{\varphi}$ is the homeomorphism between $K[X]$ and $K[Y]$ introduced in Lemma 1.5.14. See the following commutative diagram, where $y_* = y_n^2 y_{n+1} \cdots y_m$.

$$\begin{array}{ccccc}
K[X] & \xrightarrow{F_{x_m^2}} & K_{x_m^2}[X] & \xrightarrow{F_w} & K_{wx_m^2}[X] \\
\tilde{\varphi}\big\downarrow & & \big\downarrow & & \big\| \\
K[Y] & \xrightarrow[F_{y_*}]{} & K_{y_*}[X] & \xrightarrow[F_v]{} & K_{vy_*}[Y]
\end{array}$$

1.6. Rationally ramified self-similar sets 2

Continued from the last section, we will focus on rationally ramified self-similar structures. In this class, there are useful criteria for a scale being locally finite and self-similar scales being gentle with respect to each other. As in the previous sections, $\mathcal{L} = (K, S, \{F_i\}_{i \in S})$ is a self-similar structure.

THEOREM 1.6.1. *Let $\mathcal{L}$ be rationally ramified and let $\mathcal{R}$ be its relation set. Define $\mathcal{R}_2 = \{(X, Y, \varphi, x, y) \in \mathcal{R} | \#(X) \geq 2\}$. Then an elliptic scale $\mathcal{S}$ on Σ is locally finite with respect to $\mathcal{L}$ if and only if there exist $c_1, c_2 > 0$ such that*

$$c_1 l(zxx_1 \ldots x_m) \leq l(zy\varphi(x_1) \ldots \varphi(x_m)) \leq c_2 l(zxx_1 \ldots x_m) \tag{1.6.1}$$

for any $(X, Y, \varphi, x, y) \in \mathcal{R}_2$, any $x_1 \ldots x_m \in W_*(X)$ and any $z \in W_*$, where l is the gauge function of $\mathcal{S}$. In particular, for $\mathbf{a} \in (0,1)^S$, a self-similar scale $\mathcal{S}(\mathbf{a})$ on Σ is locally finite with respect to $\mathcal{L}$ if and only if $a_w = a_{\varphi(w)}$ for any $(X, Y, \varphi, x, y) \in \mathcal{R}_2$ and any $w \in X$.

COROLLARY 1.6.2. *Let $\mathcal{L}$ be rationally ramified. Assume that $\mathcal{S}_1 \cdot \mathcal{S}_2$ is locally finite with respect to $\mathcal{L}$ for elliptic scales $\mathcal{S}_1$ and $\mathcal{S}_2$ on Σ. Then $\mathcal{S}_1$ is locally finite with respect to $\mathcal{L}$ if and only if $\mathcal{S}_2$ is locally finite with respect to $\mathcal{L}$.*

To prove Theorem 1.6.1, we need the following lemma.

LEMMA 1.6.3. *Let X be a nonempty independent finite subset of $W_\#$ and let $x \in W_\#$. Assume that $O_{\Sigma[X],x}(\omega) < +\infty$ for any $\omega \in \Sigma$. Then $\sup_{w \in W_\#} \#(A_{X,x}(w)) < +\infty$.*

PROOF. By Theorem 1.5.6, we may choose k so that $Q_k(X) = \emptyset$. Fix $z_* \in W_*$. Then $(z_*, x, x_1, \ldots, x_m) \in A_{X,x}(w)$ is uniquely determined except for $x_{m-k}, \ldots, x_m$. Therefore

$$\sup_{w, z_*} \#\{(z_*, x, x_1, \ldots, x_m) \in A_{X,x}(w)\} < +\infty. \tag{1.6.2}$$

Now define

$$N_{X,x}(w) = \#\{z | (z, x, x_1, \ldots, x_m) \in A_{X,x}(w) \text{ for some } (x_1, \ldots, x_m)\}$$

Lemma 1.5.8 implies that $\sup_{\omega \in \Sigma} O_{\Sigma[X],x}(\omega) < +\infty$. Denote the value of this supremum by N. Suppose that $N_{X,x}(w) > k(N+1)$, where $k = \max_{w \in X} |w|$. For $(z, x, x_1, \ldots, x_m) \in A_{X,x}(w)$, $|w| - k \leq |zxx_1 \ldots x_m| \leq |w| - 1$. Therefore, for some $l \in \{|w| - k, \ldots, |w| - 1\}$, there exists $\{(z^{(i)}, x, x_1^{(i)}, \ldots, x_{m_i}^{(i)})\}_{i=1}^{N+1} \subset A_{X,x}(w)$

such that $z^{(i)} \neq z^{(j)}$ for any $i \neq j$ and $z^{(i)}xx_1^{(i)} \dots x_{m_i}^{(i)} = w_1 \dots w_l$, where $w = w_1 \dots w_{|w|}$. Set $\omega = w_1 \dots w_l(x_*)^\infty$, where $x_* \in X$. Then, $\sigma^{|z^{(i)}|}\omega \in \Sigma_x[X]$ for any $i = 1, \dots, N+1$. This contradicts to the definition of N. Hence, $N_{X,x}(w) \leq k(N+1)$ for any $w \in W_*$. Combining this with (1.6.2), we have the desired estimate. □

DEFINITION 1.6.4. Let $\mathcal{L}$ be a self-similar structure and let $\mathcal{R} \subseteq \mathcal{R}_{\mathcal{L}}$. For a scale $\{\Lambda_s\}_{s\in(0,1]}$, we define

$$\Lambda_{s,w}^{\mathcal{R}} = \{v|v \in \Lambda_s, \text{there exists an corresponding pair with respect to some } \Omega \in [\mathcal{R}] \text{ in } \Sigma_w \times \Sigma_v\}$$

for any $s \in (0,1]$ and any $w \in \Lambda_s$.

LEMMA 1.6.5. *Let $\mathcal{R}$ be a relation set of a rationally ramified self-similar structure $\mathcal{L}$ and let $\mathcal{S} = \{\Lambda_s\}_{s\in(0,1]}$ be a scale on $\mathcal{L}$.*
(1) *$\mathcal{S}$ is locally finite if and only if there exists $C > 0$ such that $\#(\Lambda_{s,w}^{\mathcal{R}}) \leq C$ for any $s \in (0,1]$ and any $w \in \Lambda_s$.*
(2) *Let $\psi : W_* \to [0,+\infty)$. Then, ψ is gentle with respect to $\mathcal{S}$ if and only if there exists $C' > 0$ such that $f(w) \leq C'f(v)$ for any $s \in (0,1]$, any $w \in \Lambda_s$ and any $v \in \Lambda_{s,w}^{\mathcal{R}}$.*

PROOF. Let $M = \max_{x\in K} \#(\pi^{-1}(x))$. If $v \in \Lambda_{s,w}$, then there exists $p \in K_w \cap K_v$. Choose ω and $\tau \in \pi^{-1}(p)$ so that $\omega \in \Sigma_w$ and $\tau \in \Sigma_v$. By Proposition 1.5.13-(2), we have $\omega^{(1)}, \dots, \omega^{(m+1)} \in \Sigma$ and $\Omega_1, \dots, \Omega_m \in [\mathcal{R}]$ with (AS1), (AS2) and (AS3). Hence, if $W_{s,w} = \Lambda_{s,w}^{\mathcal{R}} \cup \{w\}$,

$$\Lambda_{s,w} \subseteq \bigcup_{w^{(1)}\in W_{s,w}} \bigcup_{w^{(2)}\in W_{s,w^{(1)}}} \cdots \bigcup_{w^{(M-1)}\in W_{s,w^{(M-2)}}} W_{s,w^{M-1}}. \tag{1.6.3}$$

Now if $\#(\Lambda_{s,w}^{\mathcal{R}}) \leq C$ for any s and w, then (1.6.3) implies that $\#(\Lambda_{s,w}) \leq (C+1)^M$. Hence we have (1). Next suppose that $f(w) \leq C'f(v)$ for any $w \in \Lambda_s$ and any $v \in \Lambda_{s,w}^{\mathcal{R}}$. Then by (1.6.3), $f(w) \leq (C')^{M-1}f(v)$ for any $w \in \Lambda_s$ and any $v \in \Lambda_{s,w}$. This shows (2). □

PROOF OF THEOREM 1.6.1. Note that $\Omega \in [\mathcal{R}]$ is a finite composition of sub-relations of relations in $\mathcal{R}$ and the the number of composed sub-relations is uniformly bounded. Therefore, we may assume that $\mathcal{R} = [\mathcal{R}]$ without loss of generality. Let $\mathcal{S} = \{\Lambda_s\}_{0<s\leq 1}$. Since $\mathcal{S}$ is elliptic, there exist $\delta_1, \delta_2 \in (0,1)$ and $c > 0$ such that $(\delta_1)^{|v|}l(w) \leq l(wv) \leq c(\delta_2)^{|v|}l(w)$. Define

$$M = \max\{|w||w \in X \text{ or } w \in Y \text{ for some } (X,Y,\varphi,x,y) \in \mathcal{R}_2\}.$$

By Theorem 1.5.6, we may choose $n \geq 1$ so that $Q_n(X) = \emptyset$ for any X with $(X,Y,\varphi,x,y) \in \mathcal{R}_2$. Assume that there exist $(X,Y,\varphi,x,y) \in \mathcal{R}_2$, $x_1 \dots x_m \in W_*(X)$ and $z \in W_*$ such that $l(zxx_1 \dots x_m)(\delta_1)^{M(k+n)} > l(zyy_1 \dots y_m)$, where $y_j = \varphi(x_j)$ for $j = 1, \dots, m$. Set $s = l(zyy_1 \dots y_m)$. Then there exists $v \in W_*$ such that $v \geq zyy_1 \dots y_m$ and $v \in \Lambda_s$. Since $l(zxx_1 \dots x_{m+k+n}) > s$ for any $(x_{m+1}, \dots, x_{m+k+n}) \in X^{k+n}$, Lemma 1.5.16 implies that there exists $w \in W_*$ such that $w \leq zxx_1 \dots x_{m+k+n}$ and $w \in \Lambda_{s,v}$. Since $Q_n(X) = 0$, the set

$$\{w|x_{m+1}, \dots, x_{m+k+n} \in X, w \leq zxx_1 \dots x_{m+k+n}, w \in \Lambda_{s,v}\}$$

contains 2^k elements at least. Hence $\#(\Lambda_{s,v}) \geq 2^k$. Therefore, if $\mathcal{S}$ is locally finite with respect to $\mathcal{L}$, then we have (1.6.1) by Lemma 1.3.6.

Next we assume (1.6.1). Let $w \in \Lambda_s$ and let $\Omega = (X, Y, \varphi, x, y) \in \mathcal{R}_2$. For $\gamma = (z, x, x_1, \ldots, x_m) \in A_{X,x}(w)$, we define

$$B(\gamma) = \{(v, z, y, y_1, \ldots, y_n) | v \in \Lambda_s, (z, y, y_1, \ldots, y_n) \in A_{Y,y}(v),$$
$$x_i = \varphi(y_i) \text{ for } i = 1, \ldots, \min(m, n)\}.$$

By Lemma 1.5.16, for $v \in \Lambda^{\mathcal{R}}_{s,w}$, there exist $\gamma \in A_{X,x}(w)$ and $(v, z, y, y_1, \ldots, y_n) \in B(\gamma)$. If $\#(X) = 1$, then it is immediate to see $\#(B(\gamma)) = 1$. Suppose $\#(X) \geq 2$. Let $\gamma_* = (v, y, y_1, \ldots, y_n) \in B(\gamma)$. Since both w and v belongs to Λ_s,

$$c_3 l(zxx_1 \ldots x_m) \leq l(zyy_1 \ldots y_n) \leq c_4 l(zxx_1 \ldots x_m),$$

where c_3 and c_4 are positive constants which are independent of s, w, Ω, γ and γ_*. If $n \geq m$, then $l(zyy_1 \ldots y_n) \leq c{\delta_2}^k l(zyy_1 \ldots y_m)$, where $k = |n - m|$. Hence $c_3 l(zxx_1 \ldots x_m) \leq c(\delta_2)^k l(zyy_1 \ldots y_m)$. By (1.6.1), $|n-m|$ is bounded by a constant which is independent of s, w, Ω, γ and γ_*. (Note that the above discussion is valid even if $n < m$; we only need to exchange γ and γ_* and do the same argument.) Therefore, $\#(B(\gamma))$ is uniformly bounded with respect to w, γ. This fact with Lemma 1.6.3 implies that $\#(\Lambda^{\mathcal{R}}_{s,w})$ is uniformly bounded with respect to s and w. By Lemma 1.6.5-(1), $\mathcal{S}$ is locally finite with respect to $\mathcal{L}$.

Finally if $\mathcal{S} = \mathcal{S}(\mathbf{a})$, then it is straightforward to show that (1.6.1) is equivalent to that $a_w = a_{\varphi(w)}$ for any $(X, Y, \varphi, x, y) \in \mathcal{R}_2$ and any $w \in X$. □

For the gentleness of self-similar scales, we have the following result.

THEOREM 1.6.6. *Let $\mathcal{L} = (K, \mathcal{S}, \{F_i\}_{i \in S})$ be rationally ramified and let $\mathcal{R}$ be a relations set of $\mathcal{L}$. For $\mathbf{a} = (a_i)_{i \in S}, \mathbf{b} = (b_i)_{i \in S} \in (0,1)^S$, $\mathcal{S}(\mathbf{a}) \underset{\text{GE}}{\sim} \mathcal{S}(\mathbf{b})$ if and only if, for any $(X, Y, \varphi, x, y) \in \mathcal{R}$, either* (R1) *or* (R2) *below is satisfied:*
(R1) *$a_w = a_{\varphi(w)}$ and $b_w = b_{\varphi(w)}$ for any $w \in X$.*
(R2) *There exists $\delta > 0$ such that*

$$\delta = \frac{\log a_w}{\log b_w} = \frac{\log a_{\varphi(w)}}{\log b_{\varphi(w)}}$$

for any $w \in X$.

PROOF. We may assume that $\mathcal{R} = [\mathcal{R}]$ without loss of generality. First assume that every $\Omega \in \mathcal{R}$ satisfies (R1) or (R2). Suppose that $v \in \Lambda_{s,w}(\mathbf{a})^{\mathcal{R}}$. Then we find a corresponding pair (ω, τ) with respect to some $(X, Y, \varphi, x, y) \in \mathcal{R}$ satisfying $\omega \in \Sigma_w$, $\tau \in \Sigma_v$ and $\pi(\omega) = \pi(\tau)$. Now, let $\omega = zxx_1x_2\ldots$ and let $\tau = zyy_1y_2\ldots$, where $z \in W_*$, $x_1, x_2, \ldots \in X$ and $y_i = \varphi(x_i)$ for any i. Then we obtain that $w = zxx_1 \ldots x_n x'$ and $v = zyy_1 \ldots y_m y'$, where $x_{n+1} < x'$ and $y_{m+1} < y'$. Assume that (R2) holds. Then, $b_{y_i} = (a_{x_i})^\delta$ for any i. Now

$$\frac{b_w}{b_v} = \frac{b_x b_{x'} {a_w}^\delta (a_x a_{x'})^{-\delta}}{b_y b_{y'} {a_v}^\delta (a_y a_{y'})^{-\delta}} = \left(\frac{a_w}{a_v}\right)^\delta \left(\frac{a_y a_{y'}}{a_x a_{x'}}\right)^\delta \frac{b_x b_{x'}}{b_y b_{y'}}$$

Note that a_w/a_v is bounded (from above and below) by Lemma 1.1.13 Also since $\mathcal{R}$ and X is a finite set, $a_x, a_{x'}, b_y, b_{y'}$ is uniformly bounded. Therefore, b_w/b_v is uniformly bounded. If (R1) is satisfied, then

$$\frac{a_w}{a_v} = \frac{a_x a_{x'}}{a_y a_{y'}} a_{x_{m+1}} \ldots a_{x_n},$$

where we assume that $n \geq m$. Since a_w/a_v is uniformly bounded, it follows that $|m-n|$ is uniformly bounded from above. Hence

$$\frac{b_w}{b_v} = \frac{b_x b_{x'}}{b_y b_{y'}} b_{x_{m+1}} \dots b_{x_n}$$

is uniformly bounded (from above and below). Hence Lemma 1.6.5-(2) implies that $\mathcal{S}(\mathbf{a}) \underset{\mathrm{GE}}{\sim} \mathcal{S}(\mathbf{b})$.

Conversely assume that $\mathcal{S}(\mathbf{a}) \underset{\mathrm{GE}}{\sim} \mathcal{S}(\mathbf{b})$. Let $(X, Y, \varphi, x, y) \in \mathcal{R}$ and let $w \in X$. Since $\pi(x(w)^\infty) = \pi(y(\varphi(w))^\infty)$, Lemma 1.4.6 implies

$$\frac{\log a_w}{\log b_w} = \frac{\log a_{\varphi(w)}}{\log b_{\varphi(w)}} \tag{1.6.4}$$

We write $\delta_w = \log b_w / \log a_w$. For $x_1 \neq x_2 \in X$, write $y_i = \varphi(x_i)$ for $i = 1, 2$. Note that $\pi(x(x_1x_2)^\infty) = \pi(y(y_1y_2)^\infty)$. Hence by Lemma 1.4.6, we obtain (1.6.4) with $w = x_1x_2$. Combining the three equations (1.6.4) with $w = x_1, x_2$ and x_1x_2, we obtain either

$$\delta_{x_1} = \delta_{x_2}, \tag{1.6.5}$$

or

$$\frac{\log a_{x_1}}{\log a_{y_1}} = \frac{\log b_{x_1}}{\log b_{y_1}} = \frac{\log a_{x_2}}{\log a_{y_2}} = \frac{\log b_{x_2}}{\log b_{y_2}} \tag{1.6.6}$$

is satisfied. Suppose that (1.6.5) does not hold. Then we have (1.6.6). Write $p = \log a_{y_1} / \log a_{x_1}$. Then $a_{y_i} = (a_{x_i})^p$ and $b_{y_i} = (b_{x_i})^p$ for $i = 1, 2$. Without loss of generality, we may assume that $0 < p \leq 1$. (If not, exchange X and Y.) Suppose that $p \neq 1$. Set $x(m) = x(x_1)^m$ for any $m \geq 1$. Define $s_m = a_{x(m)} = a_x(a_{x_1})^m$. As $0 < p < 1$, for sufficiently large m, there exists a unique $n(m) \in \mathbb{N}$ such that

$$a_y(a_{y_1})^m(a_{y_2})^{n(m)-1} > s_m \geq a_y(a_{y_1})^m(a_{y_2})^{n(m)}. \tag{1.6.7}$$

Then $y(m) = y(y_1)^m(y_2)^{n(m)} \in \Lambda_{s_m}(\mathbf{a})$. Since $a_{y_i} = (a_{x_i})^p$, (1.6.7) implies that

$$n(m) - 1 \leq \frac{\log a_x - \log a_y}{\log a_{x_2}} + m\frac{(1-p)}{p}\frac{\log a_{x_1}}{\log a_{x_2}} \leq n(m). \tag{1.6.8}$$

Note that $x(m), y(m) \in \Lambda_{s_m}(\mathbf{a})$. Hence $b_{x(m)}/b_{y(m)}$ is uniformly bounded from below and above with respect to m because $\mathcal{S}(\mathbf{a}) \underset{\mathrm{GE}}{\sim} \mathcal{S}(\mathbf{b})$. Now $b_{x(m)} = b_x(a_{x_1})^{m\delta_{x_1}}$. Using Lemma 1.4.6, we obtain (1.6.4) with $w = (x_1)^m(x_2)^{n(m)}$. Therefore, if $\delta_m = \delta_{(x_1)^m(x_2)^{n(m)}}$, then $b_{y(m)} = b_y\big((a_{y_1})^m(a_{y_2})^{n(m)}\big)^{\delta_m}$. Hence,

$$\frac{b_{x(m)}}{b_{y(m)}} = \frac{b_x}{b_y}\left(\frac{a_y}{a_x}\right)^{\delta_m}\left(\frac{a_{x(m)}}{a_{y(m)}}\right)^{\delta_m}(a_{x_1})^{m(\delta_{x_1}-\delta_m)}.$$

As $\min(\delta_{x_1}, \delta_{x_2}) \leq \delta_m \leq \max(\delta_{x_1}, \delta_{x_2})$, the first three factors in the above equality is uniformly bounded from above and below with respect to m. Therefore, so is the fourth factor $(a_{x_1})^{m(\delta_{x_1}-\delta_m)}$. On the other hand, by (1.6.8),

$$\lim_{m\to\infty}(\delta_{x_1} - \delta_m) = \frac{\log_{b_{x_1}}}{\log a_{x_1}} - \frac{\log b_{x_1} + A\log b_{x_2}}{\log a_{x_1} + A\log a_{x_2}},$$

where $A = \dfrac{(1-p)}{p}\dfrac{\log a_{x_1}}{\log a_{x_2}}$. Now since $0 < p < 1$ and $\delta_{x_1} \neq \delta_{x_2}$, the value of the above limit is not zero. Therefore, $(a_{x_1})^{m(\delta_{x_1}-\delta_m)}$ is not uniformly bounded from above and below with respect to m. This contradiction implies that $p = 1$. Thus it follows that, for any $x_1 \neq x_2 \in X$, (1.6.5) holds or $a_{x_i} = a_{y_i}$ and $b_{x_i} = b_{y_i}$ for $i = 1, 2$. So if (R1) is not satisfied (i.e. there exists some $w \in X$ such that $a_w \neq a_{\varphi(w)}$ or $b_w \neq b_{\varphi(w)}$), then $\delta_w = \delta_{w'}$ for any $w' \in X$ with $w \neq w'$. This implies (R2). Thus we have the desired conclusion. □

Combining Theorems 1.6.1 and 1.6.6, we can show that the number of equivalence classes of locally finite self-similar scales under $\underset{\text{GE}}{\sim}$ is 0, 1 or $+\infty$ as follows.

THEOREM 1.6.7. *Let $\mathcal{L} = (K, S, \{F_i\}_{i \in S})$ be a rationally ramified self-similar structure and let $\mathcal{R}$ be its relation set. For any $w = w_1 \dots w_m \in W_*$, we define $f_w \in \ell(S)$ by $f_w(i) = \#(\{k | w_k = i\})$ for any $i \in S$. Define $\mathcal{R}_1 = \{(X, Y, \varphi, x, y) \in \mathcal{R} | \#(X) = 1\}$ and $\mathcal{R}_2 = \{(X, Y, \varphi, x, y) \in \mathcal{R} | \#(X) \geq 2\}$. Also let U be the subspace of $\ell(S)$ generated by $\{f_w - f_{\varphi(w)} | (X, Y, \varphi, x, y) \in \mathcal{R}_2, w \in X\}$. (If $\mathcal{R}_2 = \emptyset$, then U is thought of as $\{0\}$.)*
(1) *There exists a self-similar scale on Σ which is locally finite with respect to $\mathcal{L}$ if and only if $U \cap [0, +\infty)^S = \{0\}$.*
(2) *Assume that $U \cap [0, +\infty)^S = \{0\}$. For $\Omega = (X, Y, \varphi, x, y) \in \mathcal{R}_1$, we use U_Ω to denote the subspace of $\ell(S)$ generated by $\{f_w, f_{\varphi(w)}\}$, where $w \in X$. Also define*

$$\mathfrak{S}_{\text{LF}}(\Sigma, \mathcal{L}) = \{\mathcal{S} | \mathcal{S} \in \mathfrak{S}(\Sigma), \mathcal{S} \textit{ is locally finite with resepct to } \mathcal{L}\}.$$

If for any $\Omega \in \mathcal{R}_1$ with $\dim U_\Omega = 2$, then $\#(\mathfrak{S}_{\text{LF}}(\Sigma, \mathcal{L})/\underset{\text{GE}}{\sim}) = 1$. In other words, all self-similar scales which are locally finite with respect to $\mathcal{L}$ are gentle each other if $\dim (U \cap U_\Omega) = 1$. If $\dim (U \cap U_\Omega) = 0$ for some $\Omega \in \mathcal{R}_1$ with $\dim U_\Omega = 2$, then $\#(\mathfrak{S}_{\text{LF}}(\Sigma, \mathcal{L})/\underset{\text{GE}}{\sim}) = +\infty$.

REMARK. Let $\Omega = (X, Y, \varphi, x, y) \in \mathcal{R}_1$ and let $X = \{w\}$. Then both f_w and $f_{\varphi(w)}$ belong to $[0, +\infty)^S$. Therefore, if $U \cap [0, \infty)^S = \{0\}$ and $\dim (U \cap U_\Omega) > 0$, then $\dim (U \cap U_\Omega) = 1$.

The following lemma is a version of Stiemke's Theorem (or Minkowski-Frakas's Theorem), which is a well-known result in convex analysis. See [**39**] or [**37**] for example.

LEMMA 1.6.8. *Let X be a finite set and U be a subspace of $\ell(X)$. Then $U \cap [0, +\infty)^X = \{0\}$ if and only if $U^\perp \cap (0, +\infty)^X \neq \emptyset$, where $U^\perp$ is the orthogonal complement with respect to the inner product $(\cdot, \cdot)_X$.*

LEMMA 1.6.9. *Assume that $U \cap [0, +\infty)^S = \{0\}$. Let $\Omega = (\{w\}, \{v\}, \varphi, x, y) \in \mathcal{R}_1$. Define $\Phi_\Omega : U^\perp \to \mathbb{R}^2$ by $\Phi_\Omega(p) = \begin{pmatrix} (f_w, p)_S \\ (f_v, p)_S \end{pmatrix}$. Then $\dim \Phi_\Omega(U^\perp) = 1$ or 2. Moreover, $\dim \Phi_\Omega(U^\perp) = 1$ if and only if $\dim U_\Omega = 1$ or $\dim (U \cap U_\Omega) = 1$.*

PROOF. By Lemma 1.6.9, we have $U^\perp \cap (0, +\infty)^S \neq \emptyset$. Hence $\Phi_\Omega(U^\perp) \neq \{0\}$. Hence $\dim \Phi_\Omega(U^\perp) > 0$. Since $U^\perp \cap (0, +\infty)^S \neq \emptyset$, it follows that $\dim \Phi_\Omega(U^\perp) = 1$ if and only if there exist $\alpha > 0$ and $\beta > 0$ such that $\alpha f_w - \beta f_v \in (U^\perp)^\perp = U$. $\alpha f_w - \beta f_v = 0$ if and only if $\dim U_\Omega = 1$. Also $\alpha f_w - \beta f_v \neq 0$ if and only if $U \cap U_\Omega \neq 0$. By the remark after Theorem 1.6.7, this is equivalent to $\dim (U \cap U_\Omega) = 1$. □

PROOF OF THEOREM 1.6.7. Let $\mathcal{S} = \mathcal{S}(\mathbf{a})$, where $\mathbf{a} = (a_i)_{i \in S} \in (0, 1)^S$. Set $p_i = \log a_i$ for $i \in S$ and write $p = (p_i)_{i \in S}$. Note that $p \in (-\infty, 0)^S$. By Theorem 1.6.1, $\mathcal{S}$ is locally finite if and only if $(f_w - f_{\varphi(w)}, p)_S = 0$ for any $(X, Y, \varphi, x, y) \in$

$\mathcal{R}_2$ and any $w \in X$. This is equivalent to that $p \in U^\perp$. Therefore, there exists a self-similar scale which is locally finite if and only if $U^\perp \cap (-\infty, 0)^S \neq \emptyset$. Since U is a linear subspace, $U^\perp \cap (-\infty, 0)^S \neq \emptyset$ if and only if $U^\perp \cap (0, +\infty)^S \neq \emptyset$. Now Lemma 1.6.8 immediately implies (1).

Next we assume that $U \cap [0, +\infty)^S = \{0\}$. Let $\mathcal{S}(\mathbf{a})$ and $\mathcal{S}(\mathbf{b})$ be locally finite. By Theorem 1.6.1, the condition (R1) is satisfied for any $(X, Y, \varphi, x, y) \in \mathcal{R}_2$. If $\dim(U \cap U_\Omega) = 1$ for any $\Omega \in \mathcal{R}_1$ with $\dim U_\Omega = 2$, then Lemma 1.6.9 shows that $\dim \Phi_\Omega(U^\perp) = 1$ for any $\Omega \in \mathcal{R}_1$. This immediately implies that (R2) holds for any $\Omega \in \mathcal{R}_1$. Thus we see $\mathcal{S}(\mathbf{a}) \underset{\text{GE}}{\sim} \mathcal{S}(\mathbf{b})$. Now assume that $\dim(U \cap U_\Omega) = 0$ for some $\Omega \in \mathcal{R}_1$ with $\dim U_\Omega = 2$. Then by Lemma 1.6.9 implies that $\dim_{\Phi_\Omega(U^\perp)} = 2$. Then for any $q \in (-\infty, 0)^2$, there exists $\mathbf{a} = (a_i)_{i \in S}$ such that $\log \mathbf{a} \in U^\perp \cap (-\infty, 0)^S$ and $\Phi_\Omega(\log \mathbf{a}) = q$, where $\log \mathbf{a} = (\log a_i)_{i \in S} \in \ell(S)$. Therefore there is no constraint on the ratio between $\log a_w$ and $\log a_v$, where $\Omega = (\{w\}, \{v\}, \varphi, x, y)$. Hence $\#(\mathfrak{S}_{\text{LF}}(\Sigma, \mathcal{L})/\underset{\text{GE}}{\sim}) = +\infty$. □

COROLLARY 1.6.10. *Let $\mathcal{L}$ be rationally ramified and let $\mathcal{R}_1$ and $\mathcal{R}_2$ be the same as in Theorem 1.6.7. If $\mathcal{R}_1 = \emptyset$ and $U \cap [0, +\infty)^S = \{0\}$, then $\#(\mathfrak{S}_{\text{LF}}(\Sigma, \mathcal{L})/\underset{\text{GE}}{\sim}) = 1$.*

In the case of post critically finite self-similar structures, the above results are easy to verify as follows.

DEFINITION 1.6.11. A self-similar structure $\mathcal{L} = (K, S, \{F_i\}_{i \in S})$ is called post critically finite (p. c. f. for short) if and only if the post critical set $\mathcal{P}$ of $\mathcal{L}$ is a finite set.

PROPOSITION 1.6.12. *$\mathcal{L} = (K, S, \{F_i\}_{i \in S})$ is post critically finite if and only if $\mathcal{L}$ is rationally ramified with a relation set $\mathcal{R}$ which satisfies $\mathcal{R} = \mathcal{R}_1$. Moreover, if $\mathcal{L}$ is post critically finite, then any scale $\mathcal{S}$ of Σ is locally finite with respect to $\mathcal{L}$.*

COROLLARY 1.6.13. *Suppose that $\mathcal{L}$ is post critically finite with a relation set $\mathcal{R}$. Let*

$$\mathcal{R} = \{(\{w(j)\}, \{v(j)\}, \varphi_j, x(j), y(j)) | j = 1, \ldots, k, w(j), v(j), x(j), y(j) \in W_\#\},$$

where $\varphi_j(w(j)) = v(j)$.
(1) For $\mathbf{a} = (a_i)_{i \in S}, \mathbf{b} = (b_i)_{i \in S} \in (0, 1)^S$, $\mathcal{S}(\alpha) \underset{\text{GE}}{\sim} \mathcal{S}(\beta)$ if and only if

$$\frac{\log a_{w(j)}}{\log b_{w(j)}} = \frac{\log a_{v(j)}}{\log b_{v(j)}} \tag{1.6.9}$$

for all $i = 1, \ldots, k$.
(2) Let $\mathbf{a} = (a_i)_{i \in S} \in (0, 1)^S$. A self-similar measure μ with weight $(\mu_i)_{i \in S}$ has volume doubling property with respect to $\mathcal{S}(\mathbf{a})$ if and only if (1.6.9) with $\mathbf{b} = (\mu_i)_{i \in S}$ holds for all $j = 1, \ldots, k$.

1.7. Examples

In this section, we will apply our results in the previous sections to several examples.

EXAMPLE 1.7.1 (Sierpinski gasket). Let $(K, S, \{F_i\}_{i \in S})$ be the same as in Example 1.5.11. By Corollary 1.6.13, for $\mathbf{a} = (a_i)_{i \in S}, \mathbf{b} = (b_i)_{i \in S} \in (0, 1)^S$,

$\mathcal{S}(\mathbf{a}) \underset{\mathrm{GE}}{\sim} \mathcal{S}(\mathbf{b})$ if and only if there exists $\delta > 0$ such that $\delta = \frac{\log b_i}{\log a_i}$ for any $i \in S$. Hence $\{\mathcal{S}|\mathcal{S}$ is a self-similar scale and $\mathcal{S} \underset{\mathrm{GE}}{\sim} \mathcal{S}(\mathbf{a})\} = \{\mathcal{S}(\mathbf{a}^\delta)|\delta > 0\}$, where $\mathbf{a}^\delta = \{(a_i)^\delta\}_{i\in S}$. Also a self-similar measure μ with weight $(\mu_i)_{i\in S}$ has volume doubling property with respect to $\mathcal{S}(\mathbf{a})$ if and only if $\mu_i = (a_i)^d$, where d is the unique constant that satisfies $\sum_{i\in S}(a_i)^d = 1$.

Define

$$\mathcal{M}_{\mathrm{VD}}^{\mathrm{S}}(\mathcal{L}, \mathcal{S}) = \{(\mu)_{i\in S}|\text{the self-similar measure with weight } (\mu_i)_{i\in S} \\ \text{has volume doubling property with respect to } \mathcal{S}\}.$$

We always identify $\mathcal{M}_{\mathrm{VD}}^{\mathrm{S}}$ as the collections of self-similar measures with volume doubling property with respect to $\mathcal{S}$. For the Sierpinski gasket, $\mathcal{M}_{\mathrm{VD}}^{\mathrm{S}}(\mathcal{L}, \mathcal{S})$ consists of only one self-similar measure. In general, however, the collection of self-similar measures with volume doubling property may have richer structure. In fact, even for the Sierpinski gasket, this is the case if we change the self-similar structure.

EXAMPLE 1.7.2. $\mathcal{L} = (K, S, \{F_i\}_{i\in S})$ is the same as in Example 1.7.1. Define $\mathcal{L}_2 = (K, W_2, \{F_w\}_{w\in W_2})$. Then $\mathcal{L}_2$ is a p. c. f. self-similar structure with $\mathcal{P}_{\mathcal{L}_2} = \{(ii)^\infty|i \in S\}$. If (X, Y, φ, x, y) belongs to the relation set of $\mathcal{L}_2$, then $X = \{ii\}, Y = \{jj\}$ and $\varphi(ii) = jj$ for some $i \neq j \in S$. Let $\mathbf{a} = (a_w)_{w\in W_2} \in (0, 1)^{W_2}$ and consider $\mathcal{S}(\mathbf{a})$, the self-similar scale on $\Sigma(W_2)$ with weight $\mathbf{a}$. Also let μ be a self-similar measure with respect to $\mathcal{L}_2$ with weight $(\mu_w)_{w\in W_2}$. Then a self-similar measure μ with weight $(\mu_{ij})_{ij\in W_2}$ has the volume doubling property with respect to a self-similar scale $(a_{ij})_{ij\in W_2}$ if and only if there exists $\delta > 0$ such that $\mu_{ii} = (a_{ii})^\delta$ for any $i \in S$. In particular, if $i \neq j$, we may choose any value for μ_{ij} as long as $\sum_{w\in S}\mu_w = 1$ and $0 < \mu_{ij}$. So $\mathcal{M}_{\mathrm{VD}}^{\mathrm{S}}(K, \mathcal{S})$ is an infinite set. This fact also shows that $\mathcal{M}_{\mathrm{VD}}(\mathcal{L}, \mathcal{S})$ is not trivial for any self-similar scale $\mathcal{S}$ on $\Sigma(S)$ because $\mathcal{M}_{\mathrm{VD}}^{\mathrm{S}}(\mathcal{L}_2, \mathcal{S}_2(\mathbf{a})) \subset \mathcal{M}_{\mathrm{VD}}(\mathcal{L}, \mathcal{S}(\mathbf{a}))$, where $\mathcal{S}_2(\mathbf{a})$ is the self-similar scale on $\Sigma(W_2)$ with weight $(a_ia_j)_{ij\in W_2}$.

Next we present two examples, unit square and the Sierpinski carpet, which are not post critically finite but rationally ramified.

EXAMPLE 1.7.3 (Unit square). Let K be the unit square in $\mathbb{R}^2$, i.e. $K = [0, 1]^2$ as in Section 0.2. We will identify $\mathbb{R}^2$ with $\mathbb{C}$ is the usual manner. Let $p_1 = 0, p_2 = 1, p_3 = 1+\sqrt{-1}$ and $p_4 = \sqrt{-1}$. Define $f_i : \mathbb{C} \to \mathbb{C}$ by $f_i(x) = (x-p_i)/2+p_i$. ($\{f_i\}$'s are the same as in Section 0.2.) Then, $\mathcal{L} = (K, S, \{f_i\}_{i\in S})$, where $S = \{1, 2, 3, 4\}$, is a rationally ramified self-similar structure. To describe its relation set $\mathcal{R}$, we define $X_1 = \{1, 2\}$, $Y_1 = \{4, 3\}$, $\varphi_1(1) = 4, \varphi_1(2) = 3$, $X_2 = \{1, 4\}, Y_2 = \{2, 3\}$, $\varphi_2(1) = 2$ and $\varphi_2(4) = 3$. As we explained in Section 0.2, where φ_2 was denoted by ϕ, $(X_2, Y_2, \varphi_2, 2, 1)$ is a relation. (See Figure 0.1.) In the same way, we have a relation set

$$\mathcal{R} = \{(X_1, Y_1, \varphi_1, 4, 1), (X_1, Y_1, \varphi_1, 3, 2), (X_2, Y_2, \varphi_2, 2, 1), (X_2, Y_2, \varphi_2, 3, 4)\}.$$

Let $\mathbf{a} = (a_i)_{i\in S} \in (0, 1)^S$ and let $\mathbf{b} = (b_i)_{i\in S} \in (0, 1)^S$. Then Theorem 1.6.6 implies that $\mathcal{S}(\mathbf{a}) \underset{\mathrm{GE}}{\sim} \mathcal{S}(\mathbf{b})$ if and only if there exists $\delta > 0$ such that $\log b_i/\log a_i = \delta$ for any $i \in S$. On the other hand, by Theorem 1.6.1, $\mathcal{S}(\mathbf{a})$ is locally finite with respect to $\mathcal{L}$ if and only if $a_1 = a_2 = a_3 = a_4$. Hence,there is only one equivalence class in $\mathfrak{S}(\Sigma)/\underset{\mathrm{GE}}{\sim}$ which consists of locally finite scales. Let μ be a self-similar measure on K

and let $\mathbf{a} \in (0,1)^S$. Theorem 1.3.5 shows that μ has the volume doubling property with respect to $\mathcal{S}(\mathbf{a})$ if and only if $a_1 = a_2 = a_3 = a_4$ and μ is the restriction of the Lebesgue measure on K. So, we have only one choice of the volume doubling measure in this case. Note that if $a_i = 1/2$ for all i, $U_s(x)$ is equivalent to the Euclidean ball, i.e. there exist c_1 and c_2 such that $B_{c_1 r}(x,d) \subseteq U_r(x) \subseteq B_{c_2 r}(x,d)$ for any $r \in [0,1]$ and any $x \in K$, where d is the Euclidean distance. (In such a situation, the Euclidean distance is said to be adapted to the scale $\mathcal{S}(\mathbf{a})$. (See Section 2.3 for details.) This fact shows Theorem 0.2.1.

Changing the self-similar structure, however, we have richer structure as in the case of Sierpinski gaskets. Let $S' = \{1, \ldots, 9\}$ and let $\{F_i\}_{i \in S'}$ be the collection of contractions defined in Section 0.2. Then $\mathcal{L}' = \{K, S', \{F_i\}_{i \in S'}\}$ is a self-similar structure. Let

$$\begin{aligned}\mathcal{R} = \{(\{1,8,7\},\{3,4,5\},\psi_1,x,y)|(x,y) = (2,1),(2,3),(9,8),(4,9),(6,7),(5,6)\} \\ \cup \{(\{1,2,3\},\{7,6,5\},\psi_2,x,y)|(x,y) = (8,1),(7,8),(9,2),(6,9),(4,3),(5,4)\},\end{aligned}$$

where $\psi_1(1) = 3, \psi_1(8) = 4, \psi_1(7) = 5, \psi_2(1) = 7, \psi_2(2) = 6, \psi_2(3) = 5$. Then $\mathcal{L}'$ is rationally ramified with a relation set $\mathcal{R}$. By Theorem 1.6.1, a self-similar scale $\mathcal{S}(\mathbf{a})$ is locally finite with respect to $\mathcal{L}'$ if and only if $a_1 = a_3 = a_5 = a_7$, $a_2 = a_6$ and $a_4 = a_8$. (In Section 0.2, a ratio $\{a_i\}_{i \in S'}$ which satisfies this condition is said to be weakly symmetric.) Moreover, Corollary 1.6.10 implies that $\#(\mathfrak{S}_{\mathrm{LF}}(\Sigma,\mathcal{L})/\underset{\mathrm{GE}}{\sim}) = 1$. Combining those results with Theorem 1.3.5, we see that a self-similar measure μ with weight $\{\mu_i\}_{i \in S'}$ has the volume doubling property with respect to a self-similar scale $\mathcal{S}(\mathbf{a})$ if and only if both $\{\mu_i\}_{i \in S'}$ and $\{a_i\}_{i \in S'}$ are weakly symmetric. This fact essentially shows Theorem 0.2.3.

EXAMPLE 1.7.4 (the Sierpinski Carpet). Let $\mathcal{L}$ be the self-similar structure associated with the Sierpinski carpet given in Example 1.5.12. Fix $\mathbf{a} = (a_i)_{i \in S} \in (0,1)^S$. Using Theorem 1.6.6, we are going to determine if $\mathcal{S}(\mathbf{b}) \underset{\mathrm{GE}}{\sim} \mathcal{S}(\mathbf{a})$ holds for $\mathbf{b} = (b_i)_{i \in S}$ or not. Define two conditions (SC1) and (SC2) as follows:
(SC1) $a_1 = a_7, a_2 = a_6$ and $a_3 = a_5$
(SC2) $a_1 = a_3, a_8 = a_4$ and $a_7 = a_5$

Then there are four cases:
(A) Assume that both (SC1) and (SC2) are satisfied, i.e. $a_1 = a_3 = a_5 = a_7, a_2 = a_6$ and $a_8 = a_4$. See Figure 1.3. Then $\mathcal{S}(\mathbf{b}) \underset{\mathrm{GE}}{\sim} \mathcal{S}(\mathbf{a})$ if and only if $b_1 = b_3 = b_5 = b_7, b_2 = b_6$ and $b_8 = b_4$. So all self-similar scales on Σ with (SC1) and (SC2) are equivalent under $\underset{\mathrm{GE}}{\sim}$. Theorem 1.6.1 implies that any scale in this class is locally finite. Moreover, by Corollary 1.6.10, this is the only one equivalence class in $\mathfrak{S}(\Sigma)/\underset{\mathrm{GE}}{\sim}$ which consists of locally finite scales. Also in this case,

$$\mathcal{M}^{\mathrm{S}}_{\mathrm{VD}}(\mathcal{L},\mathcal{S}(\mathbf{a})) = \{(\mu_i)_{i \in S} | \mu_1 = \mu_3 = \mu_5 = \mu_7, \mu_2 = \mu_6, \mu_4 = \mu_8\}.$$

(B) Assume that (SC1) holds but (SC2) does not. Then $\mathcal{S}(\mathbf{b}) \underset{\mathrm{GE}}{\sim} \mathcal{S}(\mathbf{a})$ if and only if $b_1 = b_7 = (a_1)^\delta, b_3 = b_5 = (a_3)^\delta, b_4 = (a_4)^\delta, b_8 = (a_8)^\delta$ and $b_2 = b_6$ for some $\delta > 0$. In this case, as we mentioned in (A), no scale is locally finite and $\mathcal{M}_{\mathrm{VD}}(\mathcal{L},\mathcal{S}(\mathbf{a})) = \emptyset$.
(C) Assume that (SC2) holds but (SC1) does not. Then $\mathcal{S}(\mathbf{b}) \underset{\mathrm{GE}}{\sim} \mathcal{S}(\mathbf{a})$ if and only if $b_1 = b_3 = (a_1)^\delta, b_5 = b_7 = (a_5)^\delta, b_2 = (a_2)^\delta, b_6 = (a_6)^\delta$ and $b_8 = b_4$ for some

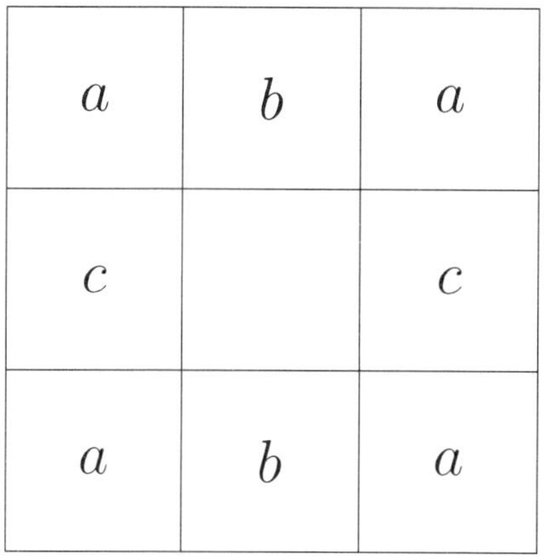

$a = a_1 = a_3 = a_5 = a_7$

$b = a_2 = a_6$

$c = a_4 = a_8$

FIGURE 1.3. Case (A) of the Sierpinski carpet

$\delta > 0$. In this case, no scale is locally finite and $\mathcal{M}_{\mathrm{VD}}(\mathcal{L}, \mathcal{S}(\mathbf{a})) = \emptyset$.
(D) Assume that neither (SC1) nor (SC2) holds. Then $\mathcal{S}(\mathbf{b}) \underset{\mathrm{GE}}{\sim} \mathcal{S}(\mathbf{a})$ if and only if there exists $\delta > 0$ such that $b_i = (a_i)^\delta$ for any $i \in S$. In this case, no scale is locally finite and $\mathcal{M}_{\mathrm{VD}}(\mathcal{L}, \mathcal{S}(\mathbf{a})) = \emptyset$.

Next we introduce a class of self-similar sets which are modifications of the Sierpinski carpet. This class contains self-similar structures which are not rationally ramified.

EXAMPLE 1.7.5 (Sierpinski cross). Let $p_1, \ldots, p_8$ be the same as in Examples 1.5.12 and1.7.4. For $r \in [1/3, 1/2)$, define

$$F_i(z) = \begin{cases} r(z - p_i) + p_i & \text{if } i \text{ is odd,} \\ (1 - 2r)(z - p_i) + p_i & \text{if } i \text{ is even.} \end{cases}$$

The unique nonempty compact set $K \subseteq \mathbb{R}^2$ satisfying $K = \cup_{i=1}^8 F_i(K)$ is called a Sierpinski cross. (Note that if $r = 1/3$, then K is the Sierpinski carpet.) Let $S = \{1, \ldots, 8\}$ and let $\mathcal{L} = (K, S, \{F_i\}_{i \in S})$. In this case, $\mathcal{L}$ may (or may not) be rationally ramified. In fact, we have the following dichotomy.

PROPOSITION 1.7.6. *Let $\mathcal{L}$ be a Sierpinski cross. Then $\mathcal{L}$ is rationally ramified if and only if r is the unique positive solution of $1 - 2r = r^m$ for some $m \in \mathbb{N}$.*

We will prove this proposition at the end of this section.

First we consider the rationally ramified cases. Assume that $1 - 2r = r^m$ for some $m \in \mathbb{N}$. Since $\mathcal{L}$ is the Sierpinski gasket for $m = 1$, we assume that $m > 1$. If $X_1, Y_1, \varphi_1, X_2, Y_2$ and φ_2 are the same as in Example 1.5.12, then the relation set

FIGURE 1.4. Sierpinski cross: rationally ramified case $r = \sqrt{2} - 1$

FIGURE 1.5. Sierpinski cross: non rationally ramified case $r = 2/5$

$\mathcal{R}$ of $\mathcal{L}$ equals

$$\{(X_l, Y_l, \varphi_l, i(j)^{m-1}, k), (X_l, Y_l, \varphi_l, k, j(i)^{m-1}) \\ |(i,j,k,l) = (7,1,8,1), (5,3,4,1), (3,1,2,2), (5,7,6,2)\}$$

Using Theorems 1.6.1 and 1.6.7, we see that $\mathcal{S}(\mathbf{a})$ is locally finite if and only if $a_1 = a_3 = a_5 = a_7$, $a_2 = a_6$ and $a_4 = a_8$. Obviously those scales are gentle each other and form an equivalence class of $\mathfrak{S}(\Sigma)/\underset{\mathrm{GE}}{\sim}$. Also a self-similar measure μ with weight $(\mu_i)_{i \in S}$ has volume doubling property with respect to those scales if and only if $\mu_1 = \mu_3 = \mu_5 = \mu_7$, $\mu_2 = \mu_6$ and $\mu_4 = \mu_8$.

Even if $\mathcal{L}$ is not rationally ramified, there exists at lease one self-similar scale on Σ that is locally finite with respect to $\mathcal{L}$. Define $\mathbf{c} = (c_i)_{i \in S}$ by $c_i = r$ if i is odd, $c_i = 1 - 2r$ if i is even. For any $w \in W_*$, define ∂K_w as the topological boundary of $F_w([0,1]^2)$. (In fact, $\partial K_w = F_w(V_0)$.) Then total length of ∂K_w is $4c_w$. Let

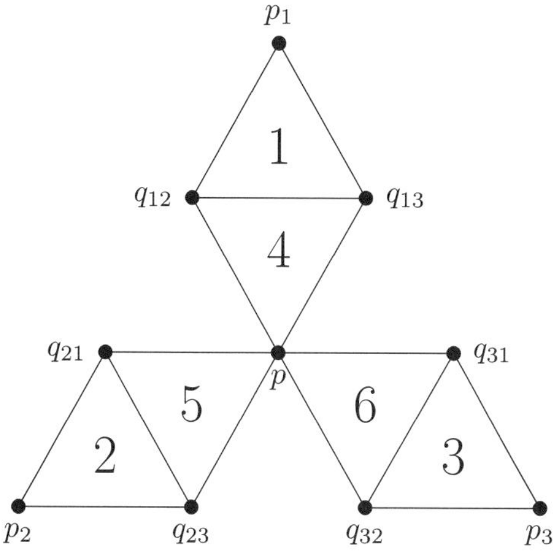

FIGURE 1.6. Construction of diamond fractal

$w \in \Lambda_s(\mathbf{c})$. Note that $c_w/(1-2r) \geq s \geq c_w$. Since $\{K_w \cap K_v\}_{v \in \Lambda_{s,w}(\mathbf{c})}$ provide a division of ∂K_w, it follows that $\#(\Lambda_s(\mathbf{c})) \leq 4(1+(1-2r)^{-1})$. Therefore, for any r, $\mathcal{S}(\mathbf{c})$ is locally finite with respect to $\mathcal{L}$.

The next example is the diamond fractal which has been introduced in [**31**]. This self-similar structure is not post critically finite but any self-similar scale is locally finite as in the post critically finite case.

EXAMPLE 1.7.7 (Diamond fractal). Let $p_1, p_2, p_3 \in \mathbb{C}$ be vertices of a regular triangle with the length of edges 1, i.e. $|p_i - p_j| = 1$ if $i \neq j$. Define $p = (p_1 + p_2 + p_3)/3$. For $i \in \{1, 2, 3\}$, define $F_i(z) = (z - p_i)/3 + p_i$ and

$$F_{i+3}(z) = -\frac{1}{3}\frac{q_i}{\overline{q_i}}(\overline{z} - \overline{p_{i+3}}) + p_{i+3},$$

where $q_i = p_i - p$ and $p_{i+3} = (3p + p_i)/4$. Let $q_{ij} = (2p_i + p_j)/3$ for any $i, j \in \{1, 2, 3\}$. If $\{i, j, k\} = \{1, 2, 3\}$, then F_{i+3} maps the regular triangle with vertices $\{p_i, p_j, p_k\}$ to the regular triangle $\{p, p_{ij}, p_{ik}\}$.

There exists a unique nonempty compact set K satisfying $K = \cup_{i=1}^{6} F_i(K)$.

K is called the diamond fractal. The corresponding self-similar structure $\mathcal{L} = (K, S, \{F_i\}_{i \in S})$, where $S = \{1, \ldots, 6\}$, is rationally ramified. In fact, the relation set $\mathcal{R}$ equals

$$\{(\{i,j\}, \{i,j\}, \mathrm{id}, k, k+3) | (i,j,k) = (1,2,3), (2,3,1), (3,1,2)\} \cup$$
$$\{(\{i\}, \{j\}, \varphi_{ij}, i+3, j+3) | (i,j) = (1,2), (2,3), (3,1)\},$$

where id is the identity map and $\varphi_{ij}(i) = j$. By Theorem 1.6.1, any self-similar scale on Σ is locally finite with respect to $\mathcal{L}$. Using Theorem 1.6.6, we see that, for $\mathbf{a}, \mathbf{b} \in (0,1)^S$, $\mathcal{S}(\mathbf{a}) \underset{\mathrm{GE}}{\sim} \mathcal{S}(\mathbf{b})$ if and only if there exists $\delta > 0$ such that $b_i = (a_i)^\delta$ for $i = 1, 2, 3$.

FIGURE 1.7. Diamond fractal

The rest of this section is devoted to a proof of Proposition 1.7.6.

LEMMA 1.7.8. *Let S be a finite set. Let $f_i : \mathbb{R} \to \mathbb{R}$ be an affine contraction for any $i \in S$, i.e. $f_i(x) = r_i x + a_i$, where $|r_i| < 1$. Let K be the self-similar set associated with $\{f_i\}_{i \in S}$. If $f_i(K) \cap f_j(K) = \emptyset$ for any $i, j \in S$ with $i \neq j$, then $\nu_1(K) = 0$, where ν_1 is the 1-dimensional Hausdorff measure.*

PROOF. Define $K_\epsilon = \{y | y \in \mathbb{R}, |x - y| \leq \epsilon \text{ for some } x \in K\}$. Let $y \in K_\epsilon$. Choose $x \in K$ so that $|x - y| \leq \epsilon$. Then $f_i(x) \in K$ and $|f(y) - f(x)| \leq |y - x| \leq \epsilon$. Hence $f_i(y) \in K_\epsilon$. This shows that $f_i(K_\epsilon) \subseteq K_\epsilon$. Let $K_\epsilon^1 = \cup_{i \in S} f_i(K_\epsilon)$. Then $K_\epsilon^1 \subseteq K_\epsilon$. Since $K_\epsilon \neq K$, the uniqueness of the self-similar set implies $K_\epsilon^1 \neq K_\epsilon$. Therefore, if $\alpha = \nu_1(K_\epsilon^1)/\nu_1(K_\epsilon)$, then $\alpha \in (0, 1)$. On the other hand, if we choose sufficiently small ϵ, then $f_i(K_\epsilon) \cap f_j(K_\epsilon) = \emptyset$ for any $i, j \in S$ with $i \neq j$. Define K_ϵ^m inductively by $K_\epsilon^m = \cup_{i \in S} f_i(K_\epsilon^{m-1})$. Then $\nu_1(K_\epsilon^m) = \alpha^m \nu_1(K_\epsilon)$. Since $K = \cap_{m \geq 0} K_\epsilon^m$, it follows that $\nu_1(K) = 0$. □

Let $\mathcal{L}$ be a Sierpinski cross.

PROOF OF PROPOSITION 1.7.6. If $1 - 2r = r^m$ for some $m \in \mathbb{R}$, then we can give the relation set $\mathcal{R}$ as in Example 1.7.5. Hence $\mathcal{L}$ is rationally ramified.

Next assume that $\mathcal{L}$ is rationally ramified with a relation set $\mathcal{R}$. Let $[\mathcal{R}] = \{\Omega_1, \ldots, \Omega_m\}$, where $\Omega = (X_i, Y_i, \varphi_i, x(i), y(i))$, be a relation set. Consider $K_8 \cap K_7 = F_7(L_1) \cap F_8(L_2) = F_8(L_2)$, where $L_1 = [0, 1]$ and $L_2 = \{x + \sqrt{-1} | x \in [0, 1]\}$. Define $J = \{i | x(i) \in \sigma_7(W_*), y(i) \in \sigma_8(W_*), \nu_1(K_{x(i)}[X_i] \cap K_7 \cap K_8) > 0\}$. By (1.5.1), $K_7 \cap K_8 \subseteq \cup_{i \in J} K_{x(i)}[X_i]$. Choose $i \in J$ so that $K_{x(i)}[X_i]$ contains $F_7(0) = F_8(\sqrt{-1})$ and write $X = X_i, Y = Y_i, \varphi = \varphi_i, x = x(i)$ and $y = y(i)$ for simplicity. Since $\pi^{-1}(F_7(0)) = \{8(7)^\infty, 7(1)^\infty\}$, we see that $x = 7(1)^p, X \subseteq W_\#(\{1, 2, 3\}), y = 8(7)^q$ and $Y \subseteq W_\#(\{7, 6, 5\})$. Note that $\nu_1(K[X]) > 0$ and that $K[X]$ is the self-similar set associated with $\{F_w\}_{w \in X}$. By Lemma 1.7.8, $F_w(K[X]) \cap F_v(K[X]) \neq \emptyset$ for some $w, v \in X$. Note that $K[X] \subseteq L_1 = [0, 1]$. The intersection $F_w(L_1) \cap F_v(L_1)$ contains a pair of points $\{\pi(w_*(1)^\infty), \pi(v_*(3)^\infty\}$ if it is not empty. Hence there exists a $w \in X$ such that $w = (3)^m$. This implies $\pi((3)^\infty) = 1 \in K[X]$. Using the same arguments, we also obtain that $\pi(5^\infty) = 1 + \sqrt{-1} \in K[Y]$. Since $1 + \sqrt{-1}$

and 1 are the most right points in $K[X]$ and $K[Y]$ respectively, $F_{8(7)^p}(1+\sqrt{-1}) = F_{7(1)^q}(1)$. Therefore, $1 - 2r = r^{q+1-p}$. Since $0 < 1 - 2r < 1$, it follows that $q+1-p \geq 1$. □

CHAPTER 2

Construction of Distances

2.1. Distances associated with scales

We have studied the scale and the associated family of "balls" $\{U_s(x)\}_{x\in X, s\in(0,1]}$ in the previous sections. Can this family of "balls"be thought of as real balls with respect to any distance? The next three sections are devoted to answer this question. In this section, we will define a pseudodistance on a self-similar set associated with a scale and consider when this pseudodistance is a distance.

As in the previous sections, S is a finite set, $\mathcal{S} = \{\Lambda_s\}_{0<s\leq 1}$ is a right-continuous scale on $\Sigma = \Sigma(S)$ whose gauge function is l and $\mathcal{L} = (K, S, \{F_i\}_{i\in S})$ is a self-similar structure. Moreover, we assume that K is connected in the following sections.

DEFINITION 2.1.1. A sequence of words, $(w(1), \ldots, w(m))$,where $w(i) \in W_*$ for any i, is called a chain of $\mathcal{L}$ if and only if $K_{w(i)} \cap K_{w(i+1)} \neq \emptyset$ for $i = 1, \ldots, m-1$. We use $\mathcal{CH}$ to denote the collection of all chains of $\mathcal{L}$. A chain $(w(1), \ldots, w(m))$ is said to be a chain between x and y for $x, y \in K$ if and only if $x \in K_{w(1)}$ and $y \in K_{w(m)}$. The collection of all chains between x and y is denoted by $\mathcal{CH}(x, y)$.

Since K is assumed to be connected, $\mathcal{CH}(x, y) \neq \emptyset$ for any $x, y \in K$. See [**28**, Theorem 1.6.2].

PROPOSITION 2.1.2. *For $x, y \in K$, we define $D_{\mathcal{S}}(x, y)$ by*

$$D_{\mathcal{S}}(x, y) = \inf\{\sum_{i=1}^{m} l(w(i)) | (w(1), \ldots, w(m)) \in \mathcal{CH}(x, y)\}.$$

Then $D_{\mathcal{S}}(\cdot, \cdot)$ is a pseudodistance on K: $D_{\mathcal{S}}(x, y) = D_{\mathcal{S}}(y, x) \geq 0$ for any x, y, $D_{\mathcal{S}}(x, x) = 0$ for any $x \in K$ and $D_{\mathcal{S}}(x, z) \leq D_{\mathcal{S}}(x, y) + D_{\mathcal{S}}(y, z)$ for any $x, y, z \in K$. Also $D_{\mathcal{S}}(F_w(x), F_w(y)) \leq l(w)$ for any $x, y \in K$. Moreover, if $D_{\mathcal{S}}(\cdot, \cdot)$ is a distance on K, then it is compatible with the original topology of K.

PROOF. It is straight forward to see that $D_{\mathcal{S}}$ is a pseudodistance on K by its definition. Since both $F_w(x)$ and $F_w(y)$ belong to K_w, $D_{\mathcal{S}}(F_w(x), F_w(y)) \leq l(w)$. Now assume that $D_{\mathcal{S}}$ is a distance on K, i.e. $D_{\mathcal{S}}(x, y) \geq 0$ for any $x, y \in K$. Note that $\{U_s(x)\}_{0<s\leq 1}$ is a system of fundamental neighborhoods of x with respect to the original topology. Let d be a distance on K which gives the original topology of K. Suppose that $d(x_n, x) \to 0$ as $n \to \infty$ for a sequence $\{x_n\}_{n\geq 1}$. Then, for any $s > 0$, x_n belongs to $U_s(x)$ for sufficiently large n. Hence $D_{\mathcal{S}}(x, x_n) \leq 2s$ for sufficiently large n. This implies that $D_{\mathcal{S}}(x, x_n) \to 0$ as $n \to \infty$. Conversely assume that $D_{\mathcal{S}}(x_n, x) \to 0$ as $n \to \infty$. Let y be an accumulating point of $\{x_n\}_{n\geq 1}$ with respect to d: there exists a subsequence $\{y_m\}_{m\geq 1}$ of $\{x_n\}_{n\geq 1}$ such that $d(y_m, y) \to 0$ as $m \to \infty$. Since $D_{\mathcal{S}}(y_m, x) \to 0$ as $m \to \infty$, we see that $D_{\mathcal{S}}(x, y) = 0$. Hence $x = y$. Now the compactness of (K, d) implies that $d(x_n, x) \to 0$ as $n \to \infty$. □

DEFINITION 2.1.3. $D_{\mathcal{S}}$ is called the pseudodistance on K associated with the scale $\mathcal{S}$. In particular, if $\mathcal{S} = \mathcal{S}(\mathbf{a})$ for $\mathbf{a} \in (0,1)^S$, then we write $D_{\mathcal{S}} = D_{\mathbf{a}}$.

REMARK. If $\mathcal{S} = \mathcal{S}(\mathbf{a})$ for $\mathbf{a} \in (0,1)^S$, then $D_{\mathcal{S}}$ coincides with the standard pseudodistance on K with poly ratio $\mathbf{a}$ defined by Kameyama [**26**].

NOTATION. Let d be a (pseudo)distance on K. For $x \in K$ and $r > 0$, we define $B_r(x,d) = \{y|y \in K, d(x,y) \leq r)$. Also $\operatorname{diam}(A,d) = \sup_{x,y \in A} d(x,y)$ for $A \subseteq K$.

$B_r(x,d)$ is the r-ball around x with respect to d and $\operatorname{diam}(A,d)$ is the diameter of A with respect to d. A ball with respect to the pseudodistance $D_{\mathcal{S}}$ always contains a "ball" associated with the scale $\mathcal{S}$ as follows.

PROPOSITION 2.1.4. *For any $n \in \mathbb{N} \cup \{0\}$, any $s \in (0,1]$ and any $x \in K$,*

$$U_s^{(n)}(x) \subseteq B_{(n+1)s}(x, D_{\mathcal{S}}),$$

PROOF. Let $y \in U_s^{(n)}(x)$. Then there exists a chain $(w(1), \ldots, w(m))$ between x and y such that $m \leq n+1$ and $w(j) \in \Lambda_s$ for any j. Since $l(w(j)) \leq s$ for any j, it follows that $\sum_{j=1}^m l(w(j)) \leq (n+1)s$. □

In general, we have the next equivalence between conditions concerning a distance and a pseudodistance associated with a scale.

PROPOSITION 2.1.5. *Let d be a distance on K and let $\beta > 0$. Then the following four conditions are equivalent.*

(1) $K_s(x) \subseteq B_{\beta s}(x,d)$ *for any $x \in K$ and any $s \in (0,1]$.*
(2) $U_s^{(n)}(x) \subseteq B_{(n+1)\beta s}(x,d)$ *for any $n \geq 0$, any $x \in K$ and any $s \in (0,1]$.*
(3) $d(x,y) \leq \beta D_{\mathcal{S}}(x,y)$ *for any $x, y \in K$.*
(4) $\operatorname{diam}(K_w, d) \leq \beta l(w)$ *for any $w \in W_*$.*

In particular, if any of the four conditions above is satisfied, $D_{\mathcal{S}}$ is a distance on K.

Recall that $K_s(x) = U_s^{(0)}(x)$.

PROOF. (1) $\Rightarrow$ (3): Let $(w(j))_{j=1,\ldots,m} \in \mathcal{CH}(x,y)$. Choose $x_j \in K_{w(j)} \cap K_{w(j+1)}$ for $j = 1, \ldots, m-1$. Set $x_0 = x$ and $x_m = y$. Then $x_j \in U_{l(w(j))}(x_{j-1}) \subseteq B_{\beta l(w(j))}(x_{j-1})$ for $j = 1, \ldots, m$. Hence $d(x_{j-1}, x_j) \leq \beta l(w(j))$. Summing these inequalities for $j = 1$ to $j = m$, we obtain $d(x,y) \leq \beta D_{\mathcal{S}}(x,y)$.
(3) $\Rightarrow$ (2): Since $B_s(x, D_{\mathcal{S}}) \subseteq B_{\beta s}(x,d)$, Proposition 2.1.4 suffices to see the claim.
(2) $\Rightarrow$ (1): Obvious
(3) $\Rightarrow$ (4): Let x and y belong to K_w. Since $D_{\mathcal{S}}(x,y) \leq l(w)$, it follows that $d(x,y) \leq \beta l(w)$.
(4) $\Rightarrow$ (1): Let $y \in U_s^{(0)}(x)$. Then $x, y \in K_w$ for some $w \in \Lambda_s$. Since $d(x,y) \leq \beta l(w)$, we obtain (1). □

If we can find one elliptic scale $\mathcal{S}_*$ where $D_{\mathcal{S}_*}$ is a distance, then for any elliptic scale $\mathcal{S}$, $D_{\mathcal{S}^\alpha}$ is a distance for some $\alpha > 0$. To give detailed version of such a result, we need the following definition.

DEFINITION 2.1.6. Let $\mathcal{S}$ be a scale on Σ and let l be its gauge function. For $w \in W_*$, define $l_w : W_* \to (0,1]$ by $l_w(v) = l(wv)/l(w)$. We denote the scale whose gauge function is l_w by $\mathcal{S}_w$.

In the above definition, it is obvious that $\mathcal{S}_w$ is actually a (right-continuous) scale. In the followings, we use $\mathcal{S}_w^\alpha$ to denote $(\mathcal{S}^\alpha)_w$ for $\alpha > 0$ and $w \in W_*$. Note that $(\mathcal{S}^\alpha)_w = (\mathcal{S}_w)^\alpha$.

PROPOSITION 2.1.7. *Let $\mathcal{S}_*$ be a scale on Σ with* (EL2)*. Suppose that $D_{\mathcal{S}_*}$ is a distance on K. If a scale $\mathcal{S}$ satisfies* (EL1)*, then there exist $\alpha > 0$ and $\beta > 0$ such that $D_{\mathcal{S}_w^\alpha}(x,y) \geq \beta D_{\mathcal{S}_*}(x,y)$ for any $x, y \in K$ and any $w \in W_*$. In particular, $D_{\mathcal{S}_w^\alpha}$ is a distance on K.*

PROOF. Let l_* be the gauge function of $\mathcal{S}_*$. By (EL2), there exists $\gamma \in (0,1)$ and $\beta > 0$ such that $l_*(v) \leq \beta\gamma^{|v|}$ for any $v \in W_*$. Also if l is the gauge function of $\mathcal{S}$, then (EL2) implies that there exists $\beta_1 \in (0,1)$ such that $l_w(v) \geq (\beta_1)^{|v|}$ for any $v, w \in W_*$. Therefore if $(\beta_1)^\alpha \geq \gamma$, then $\mathrm{diam}(K_v, D_{\mathcal{S}_*}) \leq \beta^{-1} l_w(v)$ for any $v \in W_*$. By Proposition 2.1.5, we see that $D_{\mathcal{S}_w^\alpha}(x,y) \geq \beta D_{\mathcal{S}_*}(x,y)$. □

Next theorem gives a topological sufficient condition for $D_{\mathcal{S}}$ being a distance. By virtue of this result, for any locally finite scale $\mathcal{S}$ on a rationally ramified self-similar structures, $D_{\mathcal{S}^\alpha}$ is shown to be a distance for some $\alpha > 0$ in the next section.

THEOREM 2.1.8. *Let $\mathcal{S} = \{\Lambda_s\}_{0<s\leq 1}$ be a scale on Σ. Assume the existence of $n \in \mathbb{N}$ satisfying the following two conditions* (D1) *and* (D2)*:*
(D1) *If $w \in \Lambda_s$, $\tau \in W_n$, $v \in \Lambda_{s,w}$ and $K_{v\tau} \cap K_w \neq \emptyset$, then $K_{v\tau} \cap K_{v'} = \emptyset$ for any $v' \in \Lambda_s \backslash \Lambda_{s,w}$.*
(D2) *Let l be the gauge function of $\mathcal{S}$. Set $\beta = (\sqrt{17} - 1)/4$. Then $l(w\tau) \geq \beta l(w)$ for any $w \in W_*$ and any $\tau \in W_n$.*
Then for any $x, y \in K$,

$$\inf\{s | y \in U_s^{(3)}(x)\} \leq D_{\mathcal{S}}(x,y) \leq 4\inf\{s | y \in U_s^{(3)}(x)\}.$$

In particular, $D_{\mathcal{S}}$ is a distance on K. Moreover, for any $s \in (0,1]$ and any $x \in K$,

$$B_s(x, D_{\mathcal{S}}) \subseteq U_s^{(3)}(x) \subseteq B_{4s}(x, D_{\mathcal{S}}).$$

Note that $0 < \beta < 1$.

The condition (D1) is shown to hold if $\mathcal{S}$ is intersection type finite in the next section. See the next section for the notion of "intersection type finite".

To prove the above theorem, we need several lemmas.

LEMMA 2.1.9. *For $w \in \Lambda_s$, we define $U_s(w) = K(\Lambda_{s,w}) = K(W(\Lambda_s, K_w))$. Assume that* (D1) *is satisfied. If $w \in \Lambda_s$, $\tau \in W_n$, $v \in \Lambda_{s,w}$, $v\tau \in \Lambda_{s'}$ and $K_{v\tau} \cap K_w \neq \emptyset$, then $U_{s'}(v\tau) \subseteq U_s(w)$.*

PROOF. Let $v'' \in \Lambda_{s',v\tau}$. Since $K_{v\tau} \cap K_{v'} = \emptyset$ for any $v' \in \Lambda_{s,w}$, there exists $w' \in \Lambda_{s,w}$ such that $w' \geq v''$. Therefore, $K_{v''} \subseteq K_{w'} \subseteq U_s(w)$. □

LEMMA 2.1.10. *Assume* (D1) *and* (D2)*. Let (w, v) be a chain of $\mathcal{L}$. If $w \in \Lambda_s, v \in \Lambda_{s'}$ and $\beta l(w) \geq l(v)$, then $U_{s'}(v) \subseteq U_s(w)$.*

PROOF. If $|v| \leq n$, then $l(v) \geq \beta l(\emptyset) \geq \beta l(w) \geq l(v)$. Therefore, $1 = l(\emptyset) = l(w)$. Since $w \in \Lambda_s$, we see that $w = \emptyset$ and $s = 1$. Hence $U_{s'}(v) \subseteq U_s(w) = K$. Assume that $|v| > n$. Let $v = v'z$ for $z \in W_n$. Then $l(v) \geq \beta l(v')$. This implies $l(w) \geq l(v')$. Therefore, $v = v_*\tau\tau'$ for $v_* \in \Lambda_s$, $\tau \in W_n$ and $\tau' \in W_*$. Since $K_{v_*x} \cap K_w \neq \emptyset$, Lemma 2.1.9 implies that $U_{l(v_*\tau)}(v_*x) \subseteq U_s(w)$. Note that $s' \leq l(v_*\tau)$. Hence $U_{s'}(v) \subseteq U_s(w)$. □

LEMMA 2.1.11. *Assume* (D1) *and* (D2). *Let* (v, w, τ) *be a chain of* $\mathcal{L}$. *If* $\beta l(w) < l(v)$ *and* $\beta l(w) < l(\tau)$, *then there exists a chain* (v', τ') *such that* $v' \geq v, \tau' \geq \tau$ *and* $l(v') + l(\tau') < l(v) + l(w) + l(\tau)$.

PROOF. If $|w| \leq n$, then let $v' = \tau' = \emptyset$. By (D2), $(1+2\beta)\beta \geq 2$ and $l(w) \geq \beta$. Therefore, $l(v)+l(w)+l(\tau) > (1+2\beta)\beta \geq 2 = l(v')+l(\tau')$. Hence we may assume that $|w| > n$. Let $w = w_1 \dots w_m$. Set $w_* = w_1 \dots w_{m-n}$ and define $s = l(w_*)$. If $l(v) \geq s$, then we may find $v_* \in W_*$ such that $v \geq v_*, K_{v_*} \cap K_w \neq \emptyset$ and $v_* \in \Lambda_s$. If $l(v) < s$, then there exists a unique v_* such that $v_* > v$ and $v_* \in \Lambda_s$. Also we define τ_* in the same way as v_*. Since $(1 + 2\beta)\beta \geq 2$ and $(1 + \beta)\beta \geq 1$ by (D2),

$$
\begin{aligned}
l(w) + l(v) &> (1 + \beta)\beta l(w_*) \geq l(w_*) \geq l(v_*) \\
l(w) + l(\tau) &> (1 + \beta)\beta l(w_*) \geq l(w_*) \geq l(\tau_*) \qquad (2.1.1)\\
l(v) + l(w) + l(\tau) &> (1 + 2\beta)\beta l(w_*) \geq 2l(w_*) \geq l(v_*) + l(\tau_*)
\end{aligned}
$$

Since $w_* \in \Lambda_{s,v_*}$, $K_w \cap K_{v_*} \neq \emptyset$ and $K_w \cap K_{\tau_*} \neq \emptyset$, (D1) implies that $K_{v_*} \cap K_{\tau_*} \neq \emptyset$. Define $v' = \max\{v_*, v\}$ and $\tau' = \max\{\tau_*, \tau\}$. Then by (2.1.1), (v', τ') satisfies the desired properties. □

PROOF OF THEOREM 2.1.8. Define

$$\mathcal{CH}^s(x, y) = \{(w(j))_{j=1,\dots,m} | (w(j))_{j=1,\dots,m} \in \mathcal{CH}(x, y), \min_{j=1,\dots,m} l(w(j)) \geq s\}$$

$$F(s) = \inf\{\sum_{j=1}^{m} l(w(j)) | (w(j))_{j=1,\dots,m} \in \mathcal{CH}^s(x, y)\}$$

for any $s > 0$ and any $x, y \in K$. Then $F(s)$ is monotonically decreasing and $D_{\mathcal{S}}(x, y) = \lim_{s\downarrow 0} F(s)$. Note that we may only consider chains without loops in the definition of $F(s)$. Hence there exists $(w(j))_{j=1,\dots,m} \in \mathcal{CH}^s(x, y)$ which attains the infimum. Set $s_j = l(w(j))$ and $U_j = U_{s_j}(w(j))$ for $j = 1, \dots, m$. If $1 < j < m$, then Lemma 2.1.11 implies that $\beta s_j \geq s_{j-1}$ or $\beta s_j \geq s_{j+1}$. Hence by Lemma 2.1.10, $U_{j-1} \subseteq U_j$ or $U_{j+1} \subseteq U_j$. Therefore, there exists j_* such that $1 \leq j_* \leq m + 1$ and

$$U_1 \subseteq U_2 \subseteq \dots \subseteq U_{j_*-1}, U_{j_*} \supseteq \dots \supseteq U_{m-1} \supseteq U_m.$$

Let $s_* = \max\{s_{j_*-1}, s_{j_*}\}$. Since $K_{w(j_*-1)} \cap K_{w(j_*)} \neq \emptyset$, $x \in U_1$ and $y \in U_m$, we see that $y \in U^{(3)}_{s_*}(x)$. Therefore, $F(s) = \sum_{j=1}^m s_j \geq s_* \geq \inf\{s | y \in U^{(3)}_s(x)\}$. Thus $D_{\mathcal{S}}(x, y) \geq \inf\{s | y \in U^{(3)}_s(x)\}$. On the other hand, if $y \in U^{(3)}_s(x)$, then there exists $(w(1), w(2), w(3), w(4)) \in \mathcal{CH}(x, y)$ such that $w(j) \in \Lambda_s$ for $j = 1, 2, 3, 4$. Therefore, $D_{\mathcal{S}}(x, y) \leq 4s$. Hence $D_{\mathcal{S}}(x, y) \leq 4 \inf\{s | y \in U^{(3)}_s(x)\}$.

Finally, since $\{U^{(3)}_s(x)\}_{0<s\leq 1}$ is monotonically decreasing with respect to s and $\cap_{0<s\leq 1} U^{(3)}_s(x) = \{x\}$, we see that $\inf\{s | y \in U^{(3)}_s(x)\} > 0$ if $x \neq y$. □

2.2. Intersection type

Let $\mathcal{L} = (K, S, \{F_i\}_{i\in S})$ be a self-similar structure satisfying $K \backslash \overline{V}_0 \neq \emptyset$. A scale $\mathcal{S} = \{\Lambda_s\}_{s\in(0,1]}$ is said to be intersection type finite if the topological types of $K_w \cap K_v$ for $s \in (0, 1]$ and $w, v \in \Lambda_s$ are finite. Under the assumption of a scale being intersection type finite, we can verify the conditions (D1) and (D2) in Theorem 2.1.8 and hence the associated pseudodistance is a distance for some power of the scale. See Theorem 2.2.6 for details.

First we define the notion of intersection pairs.

DEFINITION 2.2.1. (1) Define $\mathcal{IP}(\mathcal{L})$ by

$$\mathcal{IP}(\mathcal{L}) = \{(w,v)|w,v \in W_{\#}, K_w \cap K_v \neq \emptyset, \Sigma_w \cap \Sigma_v = \emptyset\}.$$

$(w,v) \in \mathcal{IP}(\mathcal{L})$ is called an intersecting pair of $\mathcal{L}$.
(2) Define

$$\mathcal{A} = \{(A,B,\varphi)|A \text{ and } B \text{ are nonempty closed subsets of } V_0 \text{ and } \varphi : A \to B \text{ is a homeomorphism between } A \text{ and } B\}.$$

There exists a natural map from $\mathcal{IP}(\mathcal{L}) \to \mathcal{A}$.

PROPOSITION 2.2.2. *Define*

$$\Phi((w,v)) = ((F_w)^{-1}(K_w \cap K_v), (F_v)^{-1}(K_w \cap K_v), (F_v)^{-1} \circ F_w|_{(F_w)^{-1}(K_w \cap K_v)})$$

for any $(w,v) \in \mathcal{IP}(\mathcal{L})$. *Then* $\Phi : \mathcal{IP}(\mathcal{L}) \to \mathcal{A}$.

The image of an intersection pair under the map Φ is called the intersection type.

DEFINITION 2.2.3. (1) Define $\mathcal{IT}(\mathcal{L}) = \Phi(\mathcal{IP}(\mathcal{L}))$. An element of $\mathcal{IT}(\mathcal{L})$ is called an intersection type of $\mathcal{L}$.
(2) Let $\mathcal{S} = \{\Lambda_s\}_{0<s\leq 1}$ be a scale on Σ. Define

$$\mathcal{IP}(\mathcal{L}, \mathcal{S}) = \{(w,v)|(w,v) \in \mathcal{IP}(\mathcal{L}), w,v \in \Lambda_s \text{ for some } s \in (0,1]\}$$

and

$$\mathcal{IT}(\mathcal{L}, \mathcal{S}) = \{\Phi((w,v))|(w,v) \in \mathcal{IP}(\mathcal{L}, \mathcal{S})\}$$

$\mathcal{S}$ is said to be intersection type finite with respect to $\mathcal{L}$ if and only if $\mathcal{IT}(\mathcal{L}, \mathcal{S})$ is a finite set.

The following proposition is straight forward by definition.

PROPOSITION 2.2.4. *Let $\mathcal{S}$ be a scale on Σ. If $\mathcal{L}$ is strongly finite and $\mathcal{S}$ is intersection type finite with respect to $\mathcal{L}$, then $\mathcal{S}$ is locally finite.*

The property of a scale being intersection type finite is preserved under the equivalence relation $\underset{\text{GE}}{\sim}$.

PROPOSITION 2.2.5. *Let $\mathcal{S}_1$ and $\mathcal{S}_2$ be elliptic scales. If $\mathcal{S}_1$ is intersection type finite and $\mathcal{S}_1 \underset{\text{GE}}{\sim} \mathcal{S}_2$, then $\mathcal{S}_2$ is intersection type finite.*

PROOF. Set $\mathcal{S}_1 = \{\Lambda_s\}_{0<s\leq 1}$ and $\mathcal{S}_2 = \{\Lambda'_s\}_{0<s\leq 1}$. Also let l be the gauge function of $\mathcal{S}_1$. Suppose that $w,v \in \Lambda'_s$, that $K_w \cap K_v \neq \emptyset$ and that $l(v) \leq l(w)$. Since $\mathcal{S}_1 \underset{\text{GE}}{\sim} \mathcal{S}_2$ and $\mathcal{S}_1$ is elliptic, there exist n (which is independent of s, w and v) such that $v = v_*\tau$ and $v_* \in \Lambda_{l(w)}$ for some $v_*, \tau \in W_*$ with $|\tau| \leq n$. Therefore, $\Phi((w,v)) \in \{\Phi((w,v_*z))||z| \leq n\}$. Note that $\{\Phi((w,v_*z))||z| \leq n\}$ only depends on $\Phi((w,v_*))$. Therefore, if $\mathcal{S}_1$ is intersection type finite, then so is $\mathcal{S}_2$. □

Now we present the first main theorem of this section.

THEOREM 2.2.6. *Let $\mathcal{S}$ be a scale on Σ with* (EL1). *If $\mathcal{S}$ is intersection type finite, then there exists $\alpha > 0$ such that $D_{\mathcal{S}^\alpha}$ is a distance on K and $B_s(x, D_{\mathcal{S}^\alpha}) \subseteq U^{(3)}_{s^{1/\alpha}}(x) \subseteq B_{4s}(x, D_{\mathcal{S}^\alpha})$ for any $s \in (0,1]$ and any $x \in K$.*

PROOF. Let $\mathcal{S} = \{\Lambda_s\}_{0<s\leq 1}$ and let l be its gauge function. First we show (D1). Since $\mathcal{S}$ is intersection type finite, there exists compact subsets $B_1, \ldots, B_m \subset K$ such that $\Phi((w,v)) = (B_i, B_j, \phi_{ij})$ for any $s \in (0,1]$, any $w \in \Lambda_s$ and any $v \in \Lambda_{s,w}$. Define $W_{k,j} = \{\tau | \tau \in W_k, K_\tau \cap B_j \neq \emptyset\}$ and $K_{k,j} = \cup_{\tau \in W_{k,j}} K_\tau$ for any j. Since $\cap_{k\geq 1} K_{k,j} = B_j$, there exists n such that $K_{n,j} \cap B_p = \emptyset$ for any $j, p \in \{1, \ldots, m\}$ with $B_j \cap B_p = \emptyset$. This implies (D1).

Now note that $\mathcal{S}^\alpha$ satisfies (D1) with the same n as $\mathcal{S}$ for any $\alpha > 0$. Since $\mathcal{S}$ satisfies (EL1), there exists $\gamma \in (0,1)$ such that $l(wv) \geq \gamma l(w)$ for any $w \in W_*$ and any $v \in W_n$. Choosing α so that $\gamma^\alpha \geq \beta = (\sqrt{17}-1)/4$, we see that $\mathcal{S}^\alpha$ satisfies (D2). Thus by Theorem 2.1.8, $D_{\mathcal{S}^\alpha}$ is a distance on K. □

The second main theorem of this section tells us that one can identify "intersection type" finite with "locally" finite in the rationally ramified case.

THEOREM 2.2.7. *Let $\mathcal{L}$ be a rationally ramified self-similar structure. Then an elliptic scale $\mathcal{S}$ on Σ is intersection type finite if and only if $\mathcal{S}$ is locally finite with respect to $\mathcal{L}$.*

PROOF. Since $\mathcal{L}$ is strongly finite, if $\mathcal{S}$ is intersection type finite, then $\mathcal{S}$ is locally finite by Proposition 2.2.4. Conversely assume that $\mathcal{S}$ is locally finite. Let $\mathcal{R}$ be the relation set of $\mathcal{L}$. We may assume that $\mathcal{R} = [\mathcal{R}]$ without loss of generality. Set $\mathcal{S} = \{\Lambda_s\}_{0<s\leq 1}$.

Let $\mathcal{R}'$ be a subset of $\mathcal{R}_{\mathcal{L}}$. For $(w,v) \in \mathcal{IP}(\mathcal{L})$, define $R(w,v,\mathcal{R}') =$

$$\begin{aligned}\{(\Omega, (z, x_0, \ldots, x_m), (z, y_0, \ldots, y_n)) | \Omega = (X, Y, \varphi, x, y) \in \mathcal{R}', \\ (z, x_0, \ldots, x_m) \in A_{X,x}(w), (z, y_0, \ldots, y_n) \in A_{Y,y}(v), \\ y_j = \varphi(x_j) \text{for } j = 1, \ldots, \min\{m, n\}\}\end{aligned}$$

Let $\eta = (\Omega, (z, x_0, \ldots, x_m), (z, y_0, \ldots, y_n)) \in R(w, v, \mathcal{R}')$ with $\Omega = (X, Y, \varphi, x, y)$. Note that $z = w_1 \ldots w_N$, where $N = \inf\{i | w_i \neq v_i\} - 1$ and that the first symbols of x and y are w_{N+1} and v_{N+1} respectively. Define $K(\eta, w), K(\eta, v)$ and $\psi_\eta : K(\eta, w) \to K(\eta, v)$ as follows. If $m \geq n$, then we set $K(\eta, w) = K_{x_m^2}[X]$, $K(\eta, v) = K_{y_n^2 y_{n+1} \cdots y_m}[Y]$ and $\psi_\eta = F_{y_n^2 y_{n+1} \cdots y_m} \circ \tilde{\varphi} \circ (F_{x_m^2})^{-1}$, where x_m^2 and y_n^2 are given in Lemma 1.5.16 and $y_j = \varphi(y_j)$ for $j = n+1, \ldots, m$. If $m < n$, then $K(\eta, w) = K_{x_m^2 x_{m+1} \cdots x_n}[X]$, $K(\eta, v) = K_{y_n^2}[Y]$ and $\psi_\eta = F_{y_n^2} \circ \tilde{\varphi} \circ (F_{x_m^2 x_{m+1} \cdots x_n})^{-1}$, where $x_j = \varphi^{-1}(y_j)$ for $j = m+1, \ldots, n$. Note that $F_w(K(\eta, w)) = K_v(K(\eta, v)) \subseteq K_w \cap K_v$ and that $\psi_\eta = F_v^{-1} \circ F_w|_{K(\eta,w)}$ by Lemma 1.5.16.

Next, we define

$$\mathcal{IP}(\mathcal{L}, \mathcal{S}, \mathcal{R}') = \{(w,v) | (w,v) \in \mathcal{IP}(\mathcal{L}, \mathcal{S}), R(w, v, \mathcal{R}') \neq \emptyset\}$$

and

$$\mathcal{IT}(\mathcal{L}, \mathcal{S}, \mathcal{R}') = \{(K(\eta, w), K(\eta, v), \psi_\eta) | (w,v) \in \mathcal{IP}(\mathcal{L}, \mathcal{S}, \mathcal{R}'), \eta \in R(w, v, \mathcal{R}')\},$$

where $\mathcal{R}'$ is a subset of $\mathcal{R}_{\mathcal{L}}$. The first step of the proof is to show that $\mathcal{IT}(\mathcal{L}, \mathcal{S}, \mathcal{R}) = \mathcal{IT}(\mathcal{L}, \mathcal{S}, \mathcal{R}_1) \cup \mathcal{IT}(\mathcal{L}, \mathcal{S}, \mathcal{R}_2)$ is a finite set, where $\mathcal{R}_1$ and $\mathcal{R}_2$ are the same as in Theorem 1.6.7. First we consider $\mathcal{IT}(\mathcal{L}, \mathcal{S}, \mathcal{R}_2)$. Let $(w,v) \in \mathcal{IP}(\mathcal{L}, \mathcal{S})$ and let $\eta = (\Omega, (z, x_0, \ldots, x_m), (z, y_0, \ldots, y_n)) \in R(w, v, \mathcal{R}_2)$ with $\Omega = (X, Y, \varphi, x, y)$. Since $\mathcal{S}$ is locally finite with respect to $\mathcal{L}$, Theorem 1.6.1 implies that $|y_{n+1} \cdots y_m|$ or $|x_{m+1} \ldots x_n|$ (depending on $m \geq n$ or $m < n$) is uniformly bounded with respect to w, v and η. Also by Lemma 1.6.3, $\#(A_{X,x}(w))$ is uniformly bounded with respect to Ω and w. Therefore, $\mathcal{IT}(\mathcal{L}, \mathcal{S}, \mathcal{R}_2)$ is finite.

Secondly, let $\eta = (\Omega, (z, x_0, \ldots, x_m), (z, y_0, \ldots, y_n)) \in R(w, v, \mathcal{R}_1)$ with $\Omega = (\{x_*\}, \{y_*\}, \varphi, x, y)$. Then $K(\eta, w) \in \{\pi(\sigma^i(x(x_*)^\infty)|i = 1, 2, \ldots\}$ and $K(\eta, v) \in \{\pi(\sigma^i(y(y_*)^\infty)|i = 1, 2, \ldots\}$. Since $\mathcal{R}_1$ is finite, $\mathcal{IT}(\mathcal{L}, \mathcal{S}, \mathcal{R}_1)$ is finite. Thus it follows that $\mathcal{IT}(\mathcal{L}, \mathcal{S}, \mathcal{R})$ is a finite set.

To proceed to the next step, we need to define $\delta_m \circ \ldots \circ \delta_1$ for $\delta_1, \ldots, \delta_m \in \mathcal{A}$. For $\delta_1 = (A_1, B_1, \varphi_1)$ and $\delta_2 = (A_2, B_2, \varphi_2) \in \mathcal{A}$, define $\delta_2 \circ \delta_1 \in \mathcal{A}$ by

$$\delta_2 \circ \delta_2 = ((\varphi_1)^{-1}(A_2 \cap B_1), \varphi_2(A_2 \cap B_1), \varphi_2 \circ \varphi_1|_{(\varphi_1)^{-1}(A_2 \cap B_1)}).$$

Then $\delta_m \circ \ldots \circ \delta_1$ is defined inductively by $\delta_m \circ (\delta_{m-1} \circ \ldots \circ \delta_1)$.

Now, let $(w, v) \in \mathcal{IP}(\mathcal{L}, \mathcal{S})$ and let $p \in K_w \cap K_v$. Choose s so that $w, v \in \Lambda_s$. Choose $\omega \in \Sigma_w \cap \pi^{-1}(p)$ and $\tau \in \Sigma_v \cap \pi^{-1}(p)$. By Proposition 1.5.13-(2), there exist $\Omega_1, \ldots, \Omega_m \in \mathcal{R}$ and $\omega^{(1)}, \ldots, \omega^{(m+1)} \in \Sigma(S)$ which satisfies (AS1), (AS2) and (AS3). For some n, $\omega^{(n)} \notin \Sigma_v$ but $\omega^{(n+1)} \in \Sigma_v$. Recall the remark after Proposition 1.5.13. Set $m_j = s(\omega^{(j)}, \tau)$. Then $m_j < |v|$ for $j = 1, \ldots, n$. Hence letting $w(j) = \omega_1^{(j)} \ldots \omega_{m_j}^{(j)} (= v_1 \ldots v_{m_j})$, then $l(w(j)) \geq l(v_1 \ldots v_{|v|-1}) > s$, where l is the gauge function of $\mathcal{S}$. We may choose $k_j > m_j$ so that $v(j) = \omega_1^{(j)} \ldots \omega_{k_j}^{(j)} \in \Lambda_s$ for any $j = 1, \ldots, n$. (We set $v(1) = w$ and $v(n+1) = v$.) Let $\Omega_j = (X_j, Y_j, \varphi_j, x(j), y(j))$. Then, $\omega^{(j)} = w(j)x(j)x_1x_2 \ldots$ for some $x_1x_2 \ldots \in \Sigma[X_j]$ and $\omega^{(j+1)} = w(j)y(j)y_1y_2 \ldots$, where $y_i = \varphi_j(x_i)$. Hence, for some M_j and N_j, $\eta_j = (\Omega_j, (w(j), x(j), x_1, \ldots, x_{M_j}), (w(j), y(j), y_1, \ldots, y_{M_j}) \in R(v(j), v(j+1), \mathcal{R})$. Define $\rho_j = (K(\eta_j, v(j)), K(\eta_j, v(j+1)), \psi_{\eta_j})$. Then $\rho_j \in \mathcal{IT}(\mathcal{L}, \mathcal{S}, \mathcal{R})$. Now, $\rho_n \circ \ldots \circ \rho_1$ gives a fraction of $\Phi((w, v))$ around $F_w^{-1}(p)$. Therefore, $\Phi((w, v))$ is a combination of elements in $\{\delta_1 \circ \ldots \circ \delta_n | n \leq \max_{p \in K} \#(\pi^{-1}(p)), \delta_i \in \mathcal{IT}(\mathcal{L}, \mathcal{S}, \mathcal{R})\}$. Since this set is finite, $\mathcal{L}$ is intersection type finite. □

Combining the last two theorems, we obtain the following fact.

COROLLARY 2.2.8. *Let $\mathcal{L}$ be a rationally ramified self-similar structure. If an elliptic scale $\mathcal{S}$ on Σ is locally finite, then $D_{\mathcal{S}^\alpha}$ is a distance on K for some $\alpha > 0$. Moreover, $B_s(x, D_{\mathcal{S}^\alpha}) \subseteq U^{(3)}_{s^{1/\alpha}}(x) \subseteq B_{4s}(x, D_{\mathcal{S}^\alpha})$ for any $s \in (0, 1]$ and any $x \in K$. In particular, if $\mathcal{L}$ is post critically finite, then, for any elliptic scale on K, $D_{\mathcal{S}^\alpha}$ is a distance on K for some $\alpha > 0$.*

PROOF. The first half is verified by using Theorems 2.2.6 and 2.2.7. About post critically finite self-similar structure, recall that any scale on K is locally finite. This suffices the conclusion. □

REMARK. In **[26]**, Kameyama has shown that there exists a self-similar scale $\alpha \in (0, 1)^S$ such that D_α is a distance on K for any critically finite self-similar set, which corresponds to post critically finite self-similar structure in our language. (His definition of self-similar sets allows that the contraction mappings are not injective.) The above corollary partially extends his result to rationally ramified case.

In the rest of this section, we will give several accounts about intersection pairs. Those results are rather technical but play important roles later.

DEFINITION 2.2.9. Let $\Gamma_i \subseteq W_*$ for $i = 1, 2$. A bijection $\psi : \Gamma_1 \to \Gamma_2$ is called an $\mathcal{L}$-isomorphism between Γ_1 and Γ_2 if the following condition is satisfied:
For $w, v \in \Gamma_1$, $(w, v) \in \mathcal{IP}(\mathcal{L})$ if and only if $(\psi(w), \psi(v)) \in \mathcal{IP}(\mathcal{L})$. If $(w, v) \in \mathcal{IP}(\mathcal{L})$ for $w, v \in \Gamma_1$, then $\Phi((w, v)) = \Phi((\psi(w), \psi(v)))$.

Γ_1 and Γ_2 are said to be $\mathcal{L}$-similar if there exists an $\mathcal{L}$-isomorphism between Γ_1 and Γ_2.

PROPOSITION 2.2.10. *Let $\Gamma_i \subseteq W_*$ for $i = 1, 2$ and let ψ be an $\mathcal{L}$-isomorphism between Γ_1 and Γ_2. Then there exists a homeomorphism ϕ between $K(\Gamma_1)$ and $K(\Gamma_2)$ such that $\phi|_{K_w} = F_{\psi(w)} \circ (F_w)^{-1}$ for any $w \in \Gamma_1$. ϕ is called the $\mathcal{L}$-similitude between $K(\Gamma_1)$ and $K(\Gamma_2)$ associated with ψ.*

For $\Gamma_1, \Gamma_2 \subseteq W_*$, we say that $\phi : K(\Gamma_1) \to K(\Gamma_2)$ is an $\mathcal{L}$-similitude between $K(\Gamma_1)$ and $K(\Gamma_2)$ if and only if there exists a $\mathcal{L}$-isomorphism ψ between Γ_1 and Γ_2 and ϕ is associated with ψ.

PROOF. Let $(w, v) \in \mathcal{IP}(\mathcal{L})$ for $w, v \in \Gamma_1$. Since $\Phi((w, v)) = \Phi((\psi(w), \psi(v)))$, it follows that $F_{\psi(w)} \circ (F_w)^{-1}$ coincides with $F_{\psi(v)} \circ (F_v)^{-1}$ on $K_w \cap K_v$. Hence if $\phi = F_{\psi(w)} \circ (F_w)^{-1}$ on K_w, then ϕ is a well-defined homeomorphism between $K(\Gamma_1)$ and $K(\Gamma_2)$. □

DEFINITION 2.2.11. Let $n \in \{0\} \cup \mathbb{N}$. For $(s_1, x_1), (s_2, x_2) \in (0, 1] \times K$, we write $(s_1, x_1) \underset{n}{\sim} (s_2, x_2)$ if and only if there exists an $\mathcal{L}$-isomorphism ψ between $\Lambda^n_{s_1,x_1}$ and $\Lambda^n_{s_2,x_2}$ such that $\psi(\Lambda^k_{s_1,x_1}) = \Lambda^k_{s_2,x_2}$ for any $k = 0, 1, \ldots, n$. We call ψ the n-isomorphism between (s_1, x_1) and (s_2, x_2).

Note that $(s_1, x_1) \underset{n}{\sim} (s_2, x_2)$ implies $(s_1, x_1) \underset{k}{\sim} (s_2, x_2)$ for any $0 \leq k \leq n$. It is easy to see that $\underset{n}{\sim}$ is an equivalence relation.

PROPOSITION 2.2.12. *The relation $\underset{n}{\sim}$ is an equivalence relation on $(0, 1] \times K$ for any $n \geq 0$.*

We can relate the notion of being intersection type finite with the number of equivalence classes under $\underset{n}{\sim}$.

THEOREM 2.2.13. *Let $\mathcal{L}$ be strongly finite. Then the following three conditions are equivalent.*
(1) $\mathcal{S}$ is intersection type finite with respect to $\mathcal{L}$.
(2) $((0, 1] \times K)/\underset{n}{\sim}$ is a finite set for any $n \in \{0\} \cup \mathbb{N}$.
(3) $((0, 1] \times K)/\underset{n}{\sim}$ is a finite set for some $n \in \{0\} \cup \mathbb{N}$.

The following fact, which is used to show the above theorem, is straight forward.

LEMMA 2.2.14. *For $(s, x) \in (0, 1] \times K$ and $n \in \{0\} \cup \mathbb{N}$, define $J^n_{s,x} : \Lambda^n_{s,x} \to \{0, 1, \ldots, n\}$ by $J^n_{s,x}(w) = \min\{k | w \in \Lambda^k_{s,x}\}$ and define $H^n_{s,x} : \Lambda^n_{s,x} \times \Lambda^n_{s,x} \to \mathcal{IT}(\mathcal{L}, \mathcal{S}) \cup \{0, 1\}$ by*

$$H^n_{s,x}(w, v) = \begin{cases} \Phi((w, v)) & \text{if } (w, v) \in \mathcal{IP}(\mathcal{L}), \\ 0 & \text{if } K_w \cap K_v = \emptyset, \\ 1 & \text{if } w = v. \end{cases}$$

Then ψ is an n-isomorphism between (s_1, x_1) and (s_2, x_2) if and only if $J_{s_1,x_1}(w) = J_{s_2,x_2}(\psi(w))$ and $H_{s_1,x_1}(w, v) = H_{s_2,x_2}(\psi(w), \psi(v))$ for any $w, v \in \Lambda^n_{s_1,x_1}$.

PROOF OF THEOREM 2.2.13. (1) $\Rightarrow$ (2): Assume that $\mathcal{S}$ is intersection type finite with respect to $\mathcal{L}$. Then by Proposition 2.2.4, $\mathcal{S}$ is locally finite with respect to $\mathcal{L}$. Hence Lemma 1.3.6 implies that $\#(\Lambda^n_{s,x})$ is uniformly bounded with respect to $(s,x) \in (0,1] \times K$. Since $\mathcal{IT}(\mathcal{L},\mathcal{S})$ is a finite set, we only have finite number of choices of $J^n_{s,x}$ and $H^n_{s,x}$ up to n-isomorphisms. Therefore by Lemma 2.2.14, $\big((0,1] \times K\big)/\underset{n}{\sim}$ is a finite set for any $n \in \{0\} \cup \mathbb{N}$.
(2) $\Rightarrow$ (3): This is obvious.
(3) $\Rightarrow$ (1): We see that $\big((0,1] \times K\big)/\underset{0}{\sim}$ is a finite set under (3). Since $\mathcal{L}$ is strongly finite, $\#(\Lambda_{s,x})$ is uniformly bounded with respect to $(s,x) \in (0,1] \times K$. Therefore if $X = \cup_{(s,x)\in(0,1]\times K} \mathrm{Im}\, H^0_{s,x}$, then Lemma 2.2.14 implies that X is a finite set. Note that $\mathcal{IT}(\mathcal{L},\mathcal{S}) \subseteq X$. Thus we have $\#(\mathcal{IT}(\mathcal{L},\mathcal{S}))$ is finite. □

2.3. Qdistances adapted to scales

As is seen in the last section, $D_{\mathcal{S}}$ is not always a distance even if a scale $\mathcal{S}$ is elliptic and locally finite. Instead we sometimes managed to show that $D_{\mathcal{S}^\alpha}$ is a distance for some $\alpha > 0$. In such a case, if $d(x,y) = (D_{\mathcal{S}^\alpha}(x,y))^{1/\alpha}$, then d has the same scaling ratio as the scale $\mathcal{S}$ but d is not a distance. Considering such a situation, we will introduce the notion of an α-qdistandce in this section.

DEFINITION 2.3.1. Let X be a set. For $\alpha > 0$, $d : X \times X \to [0,\infty)$ is called α-qdistance on X if and only if $d(x,y)^\alpha$ is a distance on X. Also d is called a qdistance on X if it is an α-qdistance on X for some $\alpha > 0$. We say d_1 and d_2 are equivalent if there exist $c_1 > 0$ and $c_2 > 0$ such that $c_1 d_1(x,y) \le d_2(x,y) \le c_2 d_1(x,y)$ for any $x,y \in X$.

REMARK. We may give more general definition of a qdistance: let $f : [0,\infty) \to [0,\infty)$ satisfy that $f(x) < f(y)$ if $x, y \in [0,\infty)$ and $x < y$, $\lim_{x\downarrow 0} f(x) = f(a)$ for any $a \in [0,\infty)$ and that $f(0) = 0$. Then $d : X \times X \to [0,\infty)$ is called f-qdistance on X if and only if $f(d(x,y))$ is a distance on X. In this paper, however, we do not need such an generality. So we restrict ourselves to the case where $f(x) = x^\alpha$ for some $\alpha > 0$.

The symbol "q" of qdistance represents the prefix "quasi". We do not use the word "quasidistance" to avoid confusion with the existent notion of quasidistance (or quasimetric) which has been defined as follows: $d : X \times X \to [0,+\infty)$ is called a C-quasidistance (or quasimetric) for $C > 0$ if and only if $d(x,y) = 0$ is equivalent to $x = y$, $d(x,y) = d(y,x)$ for any x, y and

$$d(x,z) \le C(d(x,y) + d(y,z))$$

for any x, y, z. A qdistance is is a quasidistance. (In fact, an α-qdistance is a $2^{1/\alpha - 1}$-quasidistance.) The immediate converse itself is not true. We have, however, the following modification of the converse.

PROPOSITION 2.3.2. *Let d be a quasidistance on a set X. Then d is equivalent to an α-qdistance D for some $\alpha > 0$, i.e. there exist positive constants c_1 and c_2 such that $c_1 d(x,y) \le D(x,y) \le c_2 d(x,y)$ for any $x, y \in X$.*

PROOF. This proposition is a version of [**23**, Proposition 14.5] in terms of "qdistance". □

If d is a qdistance, $\lim_{n\to\infty} d(x_n, x) = 0$ implies $\lim_{n\to\infty} d(x_n, y) = d(x, y)$ for any y. This is not the case in general for a quasidistance.

If d is an α-qdistance, then d is an α'-qdistance for any $\alpha' \in (0, \alpha]$, because $a^s \leq b^s + c^s$ for any $a, b, c, \in \mathbb{R}$ with $a \leq b + c$ and any $s \in (0, 1]$. In particular, if d is an α-qdistance for $\alpha > 1$, then d is a distance. Thus, we will consider α-qdistances for $\alpha \in (0, 1]$.

For an α-qdistance d on a set X, we always associate the topology given by the distance d^α. Also we may define Hausdorff measures and Hausdorff dimensions of subsets of X in the same manner as in the case of distance as follows.

DEFINITION 2.3.3. Let d be a qdistance on X. Then for any $A \subseteq X$, we define

$$\mathcal{H}^s_\delta(A) = \inf\{\sum_{i\geq 1} \mathrm{diam}(E_i)^s | A \subseteq \cup_{i\geq 1} E_i, \mathrm{diam}(E_i) \leq \delta\}$$

for any $\delta > 0$ and $s \geq 0$, where $\mathrm{diam}(E) = \sup_{x,y\in A} d(x, y)$. Also we define $\mathcal{H}^s(A) = \lim_{\delta\downarrow 0} \mathcal{H}^s_\delta(A)$. $\mathcal{H}^s$ is called the s-dimensional Hausdorff measure with respect to the qdistance d. Also let

$$\dim_{\mathrm{H}}(A, d) = \sup\{s|\mathcal{H}^s(A) = \infty\} = \inf\{s|\mathcal{H}^s(A) = 0\}$$

for any $A \subseteq X$. $\dim_{\mathrm{H}}(A, d)$ is called the Hausdorff dimension of A with respect to the qdistance d.

As in the case of distances, $\mathcal{H}^s$ is a complete Borel regular measure on X for any $s \geq 0$.

Hereafter in this section, S is a non-empty finite set and $\mathcal{L} = (K, S, \{F_i\}_{i\in S})$ is a self-similar structure with $K \backslash \overline{V}_0 \neq \emptyset$. Also $\mathcal{S}$ is a right-continuous scale.

DEFINITION 2.3.4. A qdistance d on K is said to be adapted to a scale $\mathcal{S}$ if and only if there exists $\beta_1, \beta_2 > 0$ and $n \in \mathbb{N}$ such that

$$B_{\beta_1 s}(x, d) \subseteq U^{(n)}_s(x) \subseteq B_{\beta_2 s}(x, d)$$

for any $x \in K$ and any $s \in (0, 1]$.

For example, the distance $D_{\mathcal{S}}$ given in Theorem 2.1.8 is adapted to the scale $\mathcal{S}$ with $n = 3, \beta_1 = 1$ and $\beta_2 = 4$.

PROPOSITION 2.3.5. *If d is a qdistance on K which is adapted to a scale $\mathcal{S}$, then the topology on K given by d is the same as the original topology of K.*

PROOF. Note that $\{U^{(n)}_s(x)\}_{0<s\leq 1}$ is a fundamental system of neighborhoods of x for any $x \in K$. This immediately imply the desired statement. □

Hereafter, we always assume that the topology of K given by a qdistance d is the same as the original topology of K.

First we give an extension of Moran-Hutchinson's theorem on the Hausdorff dimension of self-similar sets.

THEOREM 2.3.6. *Let $\mathcal{S}$ be a scale on Σ which satisfies* (EL1) *and let l be the gauge function of $\mathcal{S}$. Assume that $\mathcal{S}$ is locally finite and that there exist positive constants c_1, c_2, γ and a Borel regular measure ν on K such that $c_1 l(w)^\gamma \leq \nu(K_w) \leq c_2 l(w)^\gamma$ for any $w \in W_*$. Also assume that d is a qdistance on K which is adapted to $\mathcal{S}$. Then, there exist positive constants c_3 and c_4 such that*

$$c_3 \mathcal{H}^\gamma(A) \leq \nu(A) \leq c_4 \mathcal{H}^\gamma(A) \qquad (2.3.1)$$

for any Borel set $A \subseteq K$ and

$$c_3 r^\gamma \leq \nu(B_r(x,d)) \leq c_4 r^\gamma \tag{2.3.2}$$

for any $x \in K$ and any $r > 0$. In particular, $\dim_H(K,d) = \gamma$.

PROOF. First we show that ν is elliptic. Since $\mathcal{S}$ satisfies (EL1), there exists $\beta_1 > 0$ such that $l(wi) \geq \beta_1 l(w)$ for any $w \in W_*$ and any $i \in S$. Therefore $\nu(K_{wi}) \geq c_1 l(wi)^\gamma \geq c_1(\beta_1 l(w))^\gamma$. Hence ν is elliptic. By Theorem 1.2.4, we see that $\nu \in \mathcal{M}(K)$. This implies

$$\nu(U_s^{(n)}(x)) = \sum_{w \in \Lambda_{s,x}^n} \nu(K_w).$$

for any $s \in (0,1]$ and any $x \in K$. Since μ is elliptic and $\mathcal{S}$ is locally finite, there exists positive constants c_5 and c_6 such that

$$c_5 s^\gamma \leq \nu(U_s^{(n)}(x)) \leq c_6 s^\gamma$$

for any $s \in (0,1]$ and any $x \in K$. As d is adapted to $\mathcal{S}$, this immediately shows (2.3.2). By (2.3.2), using the mass distribution principle (i.e. Frostman's lemma, see [**28**, Lemma 1.5.5]), we conclude that there exists $c_4 > 0$ such that $\nu(A) \leq c_4 \mathcal{H}^\gamma(A)$ for any Borel set A. Next fix $w \in W_*$. For sufficiently small s, define $Z_w = \{v | v \in \Lambda_s, v \leq w\}$. Then $K_w = \cup_{v \in Z_w} K_v$. Note that there exist $c' > 0$ and $c'' > 0$ such that $\mathrm{diam}(K_v, d)^\gamma \leq c' s^\gamma \leq c'' \nu(K_v)$ for any $s \in (0,1]$ and $v \in \Lambda_s$. Therefore, $\sum_{v \in Z_w} \mathrm{diam}(K_v, d)^\gamma \leq c'' \sum_{v \in Z_w} \nu(K_v) = c'' \nu(K_w)$ because $\nu \in \mathcal{M}(K)$. Since $\max_{v \in Z_w} \mathrm{diam}((K_v, d)) \to 0$ as $s \to 0$, it follows that $\mathcal{H}^\gamma(K_w) \leq c'' \nu(K_w)$. By [**28**, Theorem 1.4.10], we obtain (2.3.1). □

In general, it is difficult to find a measure ν satisfying the assumption of the above theorem. However, if $\mathcal{S}$ is a scale induced by an elliptic measure μ, then we may let $\nu = \mu$ and have $\gamma = 1$. Also there is an obvious choice of ν and γ in the case of a self-similar scale. The following corollary corresponds to the classical Moran-Hutchinson theorem on the Hausdorff dimension of a self-similar set with the open set condition. See [**28**, Section 1.5]. Also see [**35, 24**].

COROLLARY 2.3.7. *Let $\mathbf{a} = (a_i)_{i \in S} \in (0,1)^S$. Assume that $\mathcal{S}(\mathbf{a})$ is locally finite and that d is a qdistance on K which is adapted to $\mathcal{S}(\mathbf{a})$. Then the results of Theorem 2.3.6 holds, where γ is the unique constant which satisfies $\sum_{i \in S} (a_i)^\gamma = 1$ and ν is the self-similar measure with weight $((a_i)^\gamma)_{i \in S}$.*

DEFINITION 2.3.8. (1) Let $\mathcal{S}$ be a scale on Σ. For $n \geq 1$, define

$$\delta_{\mathcal{S}}^{(n)}(x,y) = \inf\{s | y \in U_s^{(n)}(x)\}$$

for any $x, y \in K$.
(2) Let d be a qdistance. We say that d is n-adapted to $\mathcal{S}$ if and only if there exist $c_1, c_2 > 0$ such that $c_1 d(x,y) \leq \delta_{\mathcal{S}}^{(n)}(x,y) \leq c_2 d(x,y)$ for any $x, y \in K$.

Obviously, a qdistance d is adapted to $\mathcal{S}$ if and only if it is n-adapted to $\mathcal{S}$ for some $n \geq 1$. If no confusion may occur, we omit $\mathcal{S}$ in $\delta_{\mathcal{S}}^{(n)}$ and write $\delta^{(n)}$. The following proposition is immediate from the definition.

PROPOSITION 2.3.9. *Let $\mathcal{S}$ be a scale on Σ. For any $n \geq 1$ and any $x, y \in K$, $\delta^{(n)}(x,y) = \delta^{(n)}(y,x)$, $\delta^{(n)}(x,y) \geq 0$ and the equality holds if and only if $x = y$.*

LEMMA 2.3.10. *Let $\mathcal{S}$ be a scale on Σ. Fix $n \in \mathbb{N}$ and $\alpha > 0$. Then the following three conditions are equivalent:*
(A) *There exists an α-qdistance which is n-adapted to $\mathcal{S}$.*
(B) *$D_{\mathcal{S}^\alpha}$ is a distance and $(D_{\mathcal{S}^\alpha})^{1/\alpha}$ is n-adapted to $\mathcal{S}$.*
(C) *$D_{\mathcal{S}^\alpha}$ is a distance and $(D_{\mathcal{S}^\alpha})^{1/\alpha}$ is m-adapted to $\mathcal{S}$ for any $m \geq n$.*

Moreover, let d be an α-qdistance. Then d is n-adapted to $\mathcal{S}$ if and only if $(D_{\mathcal{S}^\alpha})^{1/\alpha}$ is an α-qdistance which is n-adapted to $\mathcal{S}$ and d is equivalent to $(D_{\mathcal{S}^\alpha})^{1/\alpha}$.

PROOF. (A) $\Rightarrow$ (B) Let d be an α-qdistance which is n-adapted to $\mathcal{S}$. Then d^α is a distance and there exist $c_1, c_2 > 0$ such that

$$B_{c_1 s}(x, d^\alpha) \subseteq U^{(n)}_{s^{1/\alpha}}(x) \subseteq B_{c_2 s}(x, d^\alpha)$$

for any x and any s. Applying Proposition 2.1.5, we obtain $d(x,y)^\alpha \leq \beta D_{\mathcal{S}^\alpha}(x,y)$ for any x, y, where $\beta = c_2/(n+1)$. In particular, $D_{\mathcal{S}^\alpha}$ is a distance and $B_s(x, D_{\mathcal{S}^\alpha}) \subseteq B_{\beta s}(x, d^\alpha)$. Moreover, by Proposition 2.1.4, $U^{(n)}_{s^{1/\alpha}}(x) \subseteq B_{(n+1)s}(x, D_{\mathcal{S}^\alpha})$. Hence $(D_{\mathcal{S}^\alpha})^{1/\alpha}$ is n-adapted to $\mathcal{S}$.
(B) $\Rightarrow$ (C) By Proposition 2.1.4, $U^{(m)}_{s^{1/\alpha}}(x) \subseteq B_{(m+1)s}(x, D_{\mathcal{S}^\alpha})$. This along with the fact that $U^{(n)}_s(x) \subseteq U^{(m+1)}$ shows that $(D_{\mathcal{S}^\alpha})^{1/\alpha}$ is m-adapted.
(C) $\Rightarrow$ (A) This is obvious.

The remaining statement is easily verified from the arguments in "(A) $\Rightarrow$ (B)". $\square$

THEOREM 2.3.11. *Let $\mathcal{S}$ be a scale on Σ and let $n \in \mathbb{N}$. The following six properties are equivalent:*
(A) *$\delta^{(n)}$ is a quasidistance.*
(B) *There exists a qdistance which is n-adapted to $\mathcal{S}$.*
(C) *There exists $\alpha > 0$ such that $D_{\mathcal{S}^\alpha}$ is a distance and $(D_{\mathcal{S}^\alpha})^{1/\alpha}$ is n-adapted to $\mathcal{S}$.*
(D) *There exists $\alpha > 0$ such that $D_{\mathcal{S}^\alpha}$ is a distance and $(D_{\mathcal{S}^\alpha})^{1/\alpha}$ is m-adapted to $\mathcal{S}$ for any $m \geq n$.*
(E) *For any $m \geq n$, there exists $c > 0$ such that $c\delta^{(m)}(x,y) \geq \delta^{(n)}(x,y)$ for any $x, y \in K$.*
(F) *There exists $c > 0$ such that $c\delta^{(2n+1)}(x,y) \geq \delta^{(n)}(x,y)$ for any $x, y \in K$.*

PROOF. (A) $\Rightarrow$ (B) Proposition 2.3.2 suffices this implication.
(B) $\Rightarrow$ (C) $\Rightarrow$ (D) This is immediate by Lemma 2.3.10.
(D) $\Rightarrow$ (E) Let $d = (D_{\mathcal{S}^\alpha})^{1/\alpha}$. Since d is both m and n- adapted to $\mathcal{S}$, there exist $\beta_1, \beta_2 > 0$ such that $\beta_1 \delta^{(m)}(x,y) \geq \beta_2 d(x,y) \geq \delta^{(n)}(x,y)$ for any x, y.
(E) $\Rightarrow$ (F) This is obvious.
(F) $\Rightarrow$ (A) Let x, y and z belong to K. If $t > \max \delta^{(n)}(x,y), \delta^{(n)}(y,z)$, then $y \in U^{(n)}_t(x)$ and $z \in U^{(n)}_t(y)$. Hence $x \in U^{(2n+1)}_t(z)$. This shows that $\delta^{(2n+1)}(x,z) \leq \delta^{(n)}(x,y) + \delta^{(n)}(y,z)$. By (5), $\delta^{(n)}(x,z) \leq c(\delta^{(n)}(x,y) + \delta^{(n)}(y,z))$. $\square$

By the above theorem, a qdistance which is adapted to a scale $\mathcal{S}$ is essentially $(D_{\mathcal{S}^\alpha})^{1/\alpha}$. Also, qdistances which are adapted to a scale $\mathcal{S}$ are all equivalent.

COROLLARY 2.3.12. *Let $\mathcal{S}$ be a scale on Σ.*
(1) *There exists a qdistance which is adapted to $\mathcal{S}$ if and only if $(D_{\mathcal{S}^\alpha})^{1/\alpha}$ is a α-qdistance which is adapted to $\mathcal{S}$ for some $\alpha > 0$.*
(2) *Let d be a qdistance. Then d is adapted to $\mathcal{S}$ if and only if $(D_{\mathcal{S}^\alpha})^{1/\alpha}$ is a*

α-qdistance which is adapted to $\mathcal{S}$ for some $\alpha > 0$ and d is equivalent to $(D_{\mathcal{S}^\alpha})^{1/\alpha}$.
(3) *Let d_1 be a qdistances adapted to $\mathcal{S}$. Then a qdistance d_2 is adapted to $\mathcal{S}$ if and only if d_2 is equivalent to d_1.*

By the above results, if there exists a qdistance which is adapted to $\mathcal{S}$, then

$$\begin{aligned}&\{m|\text{there exists a qdistance which is } m\text{-adapted to } \mathcal{S}\}\\ =&\{m|\delta^{(m)} \text{ is a quasidistance}\} = \{n, n+1, \ldots\}.\end{aligned}$$

Denote this n by $n_A(\mathcal{S})$. Combining Theorems 2.2.6 and Corollary 2.3.12, we have the following result on existence of an adapted qdistance for an intersection type finite scale.

THEOREM 2.3.13. *Let $\mathcal{S}$ be a scale on Σ with* (EL1). *If $\mathcal{S}$ is intersection type finite with respect to $\mathcal{L}$, then there exists a qdistance on K which is adapted to $\mathcal{S}$. Furthermore, $n_A(\mathcal{S}) \leq 3$.*

PROOF. By Theorem 2.2.6, there exists $\alpha > 0$ such that $D_{\mathcal{S}^\alpha}$ is a distance on K which is 3-adapted to $\mathcal{S}^\alpha$. Therefore, if $d = (D_{\mathcal{S}^\alpha})^{1/\alpha}$, then d is a qdistance on K which is 3-adapted to $\mathcal{S}$. □

If the self-similar structure is strongly finite, then we have slightly better result.

THEOREM 2.3.14. *Assume that the self-similar structure $\mathcal{L}$ is strongly finite. If $\mathcal{S}$ is intersection type finite and satisfies* (EL1), *then $\delta^{(1)}$ is a quasidistance. In particular, $n_A(\mathcal{S}) = 1$.*

PROOF. Let $(s, x) \in (0, 1] \times K$. For any $k \geq 0$ and $m \geq 2$, define

$$\begin{aligned}\mathcal{CH}(x, s, k, m) = \{(w(1)v(1), \ldots, w(m)v(m)) \in \mathcal{CH}|w(i) \in \Lambda_s \text{ and}\\ v(i) \in W_k \text{ for any } i = 1, \ldots, m,\ x \in K_{w(1)v(1)}\}.\end{aligned}$$

Also define

$$K_m(s, x, k) = \bigcup_{(\tau(1),\ldots,\tau(m))\in\mathcal{CH}(s,x,k,m)} \left(\bigcup_{i=1}^m K_{\tau(i)}\right)$$

Let d be a distance on K which gives the original topology of K. Then the diameter of $K_m(s, x, k)$ with respect to d converges to 0 as $k \to \infty$. Since $U_s(x)$ is a neighborhood of x, there exists k_0 such that $K_m(s, x, k_0) \subseteq U_s(x)$. Since $\mathcal{S}$ satisfies (EL1), there exists $\alpha_1 \in (0, 1)$ such that $\Lambda_s \cap \Lambda_{\alpha_1 s} = \emptyset$. This means that any $w \in \Lambda_{\alpha_1 s}$ can be written as $w = w'v$, where $w' \in \Lambda_s$ and $|v| \geq 1$. Hence if $\beta = (\alpha_1)^{k_0}$, then any $w \in \Lambda_{\beta s}$ can be written as $w = w'v$, where $w' \in \Lambda_s$ and $|v| \geq k$. This along with that fact that $K_m(s, x, k_0) \subseteq U_s(x)$ yields that $U^{m-1}_{\beta s}(x) \subseteq U_s(x)$. Note that the constant β is determined by (s, x) and m. In this sense, we write $\beta = \beta(s, x, m)$.

By Theorem 2.2.13, $((0, 1] \times K)/\underset{1}{\sim}$ is a finite set. Suppose that $(s_1, x_1) \underset{1}{\sim} (s_2, x_2)$. Then there exists an $\mathcal{L}$-isomorphism ψ between $\Lambda^1_{s_1,x_1}$ and $\Lambda^1_{s_2,x_2}$. Using ψ, we see that $\beta(s_1, x_1, m) = \beta(s_2, x_2, m)$. Since the equivalence class under $\underset{1}{\sim}$ is finite, we may choose $\beta_1 \in (0, 1)$ such that $U^{(m-1)}_{\beta_1 s}(x) \subseteq U^{(n)}_s(x)$ for any $(s, x) \in (0, 1] \times K$. This implies that $\delta^{(m-1)}(x, y) \geq \beta_1 \delta^{(n)}(x, y)$ for any $x, y \in K$. In the case, $m = 4$, we have the condition (F) of Theorem 2.3.11 with $n = 1$. Therefore $\delta^{(1)}$ is a quasidistance. □

In the case of a rationally ramified self-similar structure, Corollary 2.2.8 along with Theorem 2.2.6 implies the following result.

COROLLARY 2.3.15. *Let $\mathcal{L}$ be a rationally ramified self-similar structure and let $\mathcal{S}$ be an elliptic scale on $\mathcal{S}$. If $\mathcal{S}$ is locally finite, then $n_A(\mathcal{S}) = 1$ and there exists a qdistance on K which is 1-adapted to $\mathcal{S}$. In particular, if $\mathcal{L}$ is post critically finite, then there exists an adapted qdistance for every elliptic scale on Σ.*

For self-similar scales, we have the following stronger result.

THEOREM 2.3.16. *Assume that $D_{\mathcal{S}(\mathbf{a})}$ is a distance on K, where $\mathbf{a} \in (0,1)^S$. If $\mathcal{L}$ is strongly finite and $\mathcal{S}(\mathbf{a})$ is intersection type finite, then there exists $\beta_1 > 0$ such that*

$$B_{\beta_1 s}(x, D_{\mathcal{S}(\mathbf{a})}) \subseteq U_s(x) \subseteq B_{2s}(x, D_{\mathcal{S}(\mathbf{a})})$$

for any $s \in (0,1]$ and any $x \in K$.

PROOF. By Proposition 2.1.4, we have $U_s(x) \subseteq B_{2s}(x, D_{\mathcal{S}(\mathbf{a})})$. Hereafter we write $\mathcal{S} = \mathcal{S}(\mathbf{a})$. Let $\mathbf{a} = (a_i)_{i\in S}$ and define $c = \min_{i\in S} a_i$. Let X be a finite subset of W_*. If $\cup_{w\in X} K_w$ is connected, we define, for $x, y \in \cup_{w\in X} K_w$,

$$\mathcal{CH}(x, y : X) = \{(w(1)v(1), \ldots, w(m)v(m)) \in \mathcal{CH}(x,y) | w(1), \ldots, w(m) \in X\}.$$

and $D_{\mathcal{S},X}(x,y) = \inf\{\sum_{j=1}^m a_{\tau(j)} | (\tau(1), \ldots, \tau(m)) \in \mathcal{CH}(x, y : X)\}$. Note that $D_{\mathcal{S},X}(x,y) \geq D_{\mathcal{S}}(x,y)$. Also for $(s,x) \in (0,1] \times K$, we define

$$d_{s,x} = \inf\{D_{\mathcal{S},\Lambda^2_{s,x}}(x_1, x_2) | x_1 \in K_s(x), x_2 \in U_s^{(2)}(x) \backslash U_s(x)\}$$

$$D_{s,x} = \inf\{D_{\mathcal{S}}(x_1, x_2) | x_1 \in K_s(x), x_2 \in U_s^{(2)}(x) \backslash U_s(x)\}.$$

By Theorem 2.2.13, $((0,1] \times K)/\underset{2}{\sim}$ is a finite set. Choose one representative (s_*, x_*) in a equivalence class. Suppose that $(s,x) \underset{2}{\sim} (s_*, x_*)$. Let ψ be an 2-isomorphism between (s,x) and (s_*, x_*) and let ϕ be the homeomorphism between $U_s^{(2)}(x)$ and $U_{s_*}^{(2)}(x_*)$ associated with ψ. For $p, q \in U_s^{(2)}(x)$,

$$(s_*)^{-1} \sum_{j=1}^m a_{\psi(w(j))v(j)} \leq \sum_{j=1}^m a_{v(j)} \leq (cs)^{-1} \sum_{j=1}^m a_{w(j)v(j)}$$

for any $(w(1)v(1), \ldots, w(m)v(m)) \in \mathcal{CH}(p, q : \Lambda^2_{s,x})$, where $w(1), \ldots, w(m) \in \Lambda^2_{s,x}$. Hence we have $csD_{\mathcal{S},\Lambda^2_{s_*,x_*}}(\phi(p), \phi(q))/s_* \leq D_{\mathcal{S},\Lambda^2_{s,x}}(p,q)$. This implies that $d_{s,x} \geq c_* s$, where $c_* = c(s_*)^{-1} d_{s_*,x_*}$. Since the number of equivalence classes is finite, there exists $\beta > 0$ such that $d_{s,x} \geq \beta s$ for any $(s,x) \in (0,1] \times K$.

If $d_{s,x} > D_{s,x}$, then there exists a chain between x and y which gives the infimum of the definition of $D_{s,x}$. This chain should contain a word in $\Lambda_{s'}$ for $s' \geq s$. Therefore $D_{s,x} \geq s$. Combining this with the fact that $d_{s,x} \geq \beta s$, we see that $B_{\beta_1 s}(x, D_{\mathcal{S}}) \subseteq U_s(x)$, where $\beta_1 = \beta/2$. □

Finally we define the notion of "volume doubling with respect to a qdistance" and consider measures which have volume doubling property.

THEOREM 2.3.17. *Let $\mathcal{L}$ be a rationally ramified self-similar structure and let $\mathcal{S}$ be an elliptic scale on Σ. Also let $\mu \in \mathcal{M}(K)$. Then μ has the volume doubling property with respect to $\mathcal{S}$ (i.e. (VD) is satisfied) if and only if the following condition (VDd) is satisfied:*

(VDd) *There exist a qdistance d on K which is adapted to $\mathcal{S}$, $\alpha \in (0,1)$ and $c > 0$ such that $\mu(B_s(x,d)) \leq c\mu(B_{\alpha s}(x,d))$ for any $s \in (0,1]$ and any $x \in K$.*

PROOF. If (VD) holds, then $\mathcal{S}$ is locally finite by Theorem 1.3.5. Hence by Corollary 2.3.15, there exists a qdistance on K which is adapted to $\mathcal{S}$. Now (VD) immediately implies (VDd). Conversely (VDd) implies $(\mathrm{VD})_{\mathrm{n}}$ for some n. Hence we obtain (VD). □

CHAPTER 3

Heat Kernel and Volume Doubling Property of Measures

3.1. Dirichlet forms on self-similar sets

We now begin to study heat kernels derived from "self-similar" Dirichlet forms on self-similar sets. More precisely, we will establish an equivalence between certain type of upper heat kernel estimate and the volume doubling property. See the next section for details. In this section, we will give a framework on "self-similar" Dirichlet forms. Let $\mathcal{L} = (K, S, \{F_i\}_{i\in S})$ be a self-similar structure. Hereafter we will always assume that $K \neq \overline{V}_0$ and that K is connected.

The following lemma is easy to verify.

LEMMA 3.1.1. *Let μ be an elliptic probability measure on K. Then, for any $w \in W_*$, there exists a unique elliptic probability measure μ^w on K such that $\mu^w(A) = \mu(F_w(A))/\mu(K_w)$ for any Borel set $A \subseteq K$. Moreover, define $\rho_w : L^2(K,\mu) \to L^2(K,\mu^w)$ by $\rho_w u = u\circ F_w$. Then ρ_w is a bounded operator.*

REMARK. If μ is a self-similar measure on K with weight $(\mu_i)_{i\in S}$, then $\mu^w = \mu$ for any $w \in W_*$

Now we define the notion of self-similar Dirichlet forms.

DEFINITION 3.1.2. Let μ be an elliptic probability measure on K and let $(\mathcal{E}, \mathcal{F})$ be a local regular Dirichlet form on $L^2(K,\mu)$.
(1) We say that $(\mathcal{E}, \mathcal{F}, \mu)$ is self-similar, (SSF) for short, if and only if it satisfies the following two conditions:
(SSF1) $u \circ F_i \in \mathcal{F}$ for any $i \in S$ and any $u \in \mathcal{F}$. There exists $(r_i)_{i\in S} \in (0,\infty)^S$ such that

$$\mathcal{E}(u,v) = \sum_{i\in S} \frac{1}{r_i}\mathcal{E}(u\circ F_i, v\circ F_i) \tag{3.1.1}$$

for any $u, v \in \mathcal{F}$. If $g(w) = \sqrt{r_w\mu(K_w)}$, then $g(w)$ is a gauge function and the scale $\mathcal{S}_*$ induced by g is elliptic.
(SSF2) Let Γ_1 and Γ_2 be subsets of W_* which are $\mathcal{L}$-similar and let ψ be the associated $\mathcal{L}$-similitude between $K(\Gamma_1)$ and $K(\Gamma_2)$. If $u \in \mathcal{F}$, $\mathrm{supp}(u) \subseteq K(\Gamma_1)$ and $u\circ\psi|_{\partial K(\Gamma_2)} \equiv 0$, then there exists $v \in \mathcal{F}$ such that $\mathrm{supp}(v) \subseteq K(\Gamma_2)$ and $v|_{K(\Gamma_2)} = u\circ\psi$.

The ratio $(r_i)_{i\in S}$ is called the resistance scaling ratio. If $r_i < 1$ for any $i \in S$, then $(\mathcal{E}, \mathcal{F}, \mu)$ is said to be recurrent.

REMARK. (1) If μ is a self-similar measure with weight $(\mu_i)_{i\in S}$, then $g(w)$ is a gauge function if and only if $r_i\mu_i < 1$ for any $i \in S$. In this case $\mathcal{S}_*$ is always elliptic.

(2) If $(\mathcal{E}, \mathcal{F}, \mu)$ is recurrent, then $g(w)$ is always a gauge function and $\mathcal{S}_*$ is always elliptic.

DEFINITION 3.1.3. Let μ be an elliptic probability measure on K and let $(\mathcal{E}, \mathcal{F})$ be a local regular Dirichlet form on $L^2(K, \mu)$. We say that $(\mathcal{E}, \mathcal{F}, \mu)$ satisfy Poincaré inequality, (PI) for short, if and only if there exists $c > 0$ such that

$$\text{(PI)} \qquad \mathcal{E}(u, u) \geq c \int_K (u - (\bar{u})_{\mu^w}))^2 d\mu^w$$

for any $w \in W_*$ and any $u \in \rho_w(\mathcal{F})$, where $(\bar{u})_\nu = \int_K u d\nu$.

REMARK. If μ is a self-similar measure, then $\mu^w = \mu$ for any $w \in W_*$. Therefore, in this case, (PI) holds if and only if

$$\mathcal{E}(u, u) \geq c \int_K (u - \bar{u})^2 d\mu$$

for any $u \in \mathcal{F}$, where c is a positive constant. Furthermore assume that $(\mathcal{E}, \mathcal{F})$ is conservative, i.e. $1 \in \mathcal{F}$ and $\mathcal{E}(1, 1) = 0$. Let $-\Delta$ be the non-negative self-adjoint operator associated with the Dirichlet form $(\mathcal{E}, \mathcal{F})$ on $L^2(K, \mu)$. Then by the variational principle, (PI) holds if and only if 0 is the eigenvalue of H whose multiplicity is one and the spectrum of $-\Delta$ is contained in $\{0\} \cup [c, \infty)$ for some $c > 0$.

Hereafter we always assume that μ is an elliptic probability measure on (K, d) and that $(\mathcal{E}, \mathcal{F})$ is a local regular Dirichlet form on $L^2(K, \mu)$. From the self-similarity (SSF) and the Poincaré inequality (PI), we can establish the existence of heat kernels and their diagonal estimates.

THEOREM 3.1.4. *Assume that $(\mathcal{E}, \mathcal{F}, \mu)$ satisfy the conditions* (SSF) *and* (PI). *Let $\{T_t\}_{t>0}$ be the strongly continuous semigroup on $L^2(K, \mu)$ associated with the Dirichlet form $(\mathcal{E}, \mathcal{F})$. Then $\{T_t\}_{t>0}$ is ultracontractive and there exist $\alpha > 0$ and $c > 0$ such that $||T_t||_{1\to\infty} \leq ct^{-\alpha/2}$ for any $t \in (0, 1]$. Moreover, there exists $p : (0, +\infty) \times K \times K \to [0, +\infty)$ such that $p(t, \cdot, \cdot) \in L^\infty(K \times K)$ and*

$$(T_t u)(x) = \int_K p(t, x, y) u(y) \mu(dy)$$

for any $u \in L^2(K, \mu)$. $p(t, x, y)$ is called the heat kernel associated with the Dirichlet form $(\mathcal{E}, \mathcal{F})$ on $L^2(K, \mu)$. In particular, if $(\mathcal{E}, \mathcal{F}, \mu)$ is recurrent, then $\alpha \in (0, 2)$.

We need the next two lemmas to show the above theorem.

LEMMA 3.1.5. *Let Λ be a partition of Σ. For any $u \in \mathcal{F}$, define $\Lambda(u) = \{w | w \in \Lambda, K_w \cap \mathrm{supp}(u) \neq \emptyset\}$. Assume that $(\mathcal{E}, \mathcal{F}, \mu)$ satisfy the conditions* (SSF) *and* (PI). *Then*

$$\mathcal{E}(u, u) + \frac{c}{\min_{w \in \Lambda(u)} r_w \mu(K_w)^2} ||u||_1^2 \geq \frac{c}{\max_{w \in \Lambda(u)} r_w \mu(K_w)} ||u||_2^2,$$

where c is the constant appearing in (PI).

PROOF. Using (3.1.1) and (PI), we see that

$$\begin{aligned}\mathcal{E}(u,u) &= \sum_{w\in\Lambda(u)} \frac{1}{r_w}\mathcal{E}(u\circ F_w, u\circ F_w)\\ &\geq \sum_{w\in\Lambda(u)} \frac{c}{r_w}\Big(\int_K (u\circ F_w)^2 d\mu^w - \big(\int_K u\circ F_w d\mu^w\big)^2\Big)\\ &\geq \sum_{w\in\Lambda(u)} \frac{c}{r_w\mu(K_w)}\int_{K_w} u^2 d\mu - \sum_{w\in\Lambda(u)} \frac{c}{r_w\mu(K_w)^2}\big(\int_{K_w} u d\mu\big)^2\\ &\geq \frac{c}{\max_{w\in\Lambda(u)} r_w\mu(K_w)}||u||_2^2 - \frac{c}{\min_{w\in\Lambda(u)} r_w\mu(K_w)^2}||u||_1^2.\end{aligned}$$

□

LEMMA 3.1.6. *Assume that* $(\mathcal{E},\mathcal{F},\mu)$ *satisfy the conditions* (SSF) *and* (PI). *Let* $\mathcal{S}_* = \{\Lambda_s\}_{s\in(0,1]}$. *Then there exist positive constants* c_1 *and* c_2 *such that*

$$\mathcal{E}(u,u) + \frac{c_1}{s^2\min_{w\in\Lambda_s(u)}\mu(K_w)}||u||_1^2 \geq \frac{c_2}{s^2}||u||_2^2 \tag{3.1.2}$$

for any $u\in\mathcal{F}$ *and any* $s\in(0,1]$.

PROOF. Since μ is elliptic, $\mu(K_{w_1\dots w_m}) \geq \alpha\mu(K_{w_1\dots w_{m-1}})$ for any $w_1\dots w_m \in W_*$, where $\alpha > 0$ is independent of w. Hence for any $w\in\Lambda_s$, it follows that $\alpha r s^2 \leq g(w) \leq s^2$, where $r = \min_{i\in S} r_i$. This along with Lemma 3.1.5 immediately implies (3.1.2). □

PROOF OF THEOREM 3.1.4. Since μ and $\mathcal{S}_*$ is elliptic, there exist $\delta,\eta\in(0,1)$, $c_1 > 0$ and $c_2 > 0$ such that $\mu(K_w) \geq c_1\delta^{|w|}$ and $g(w) \leq c_2\eta^{|w|}$ for any $w\in W_*$. Therefore, there exist positive constants α and c_3 such that $\mu(K_w) \geq c_3 g(w)^\alpha$ for any $w\in W_*$. This with (3.1.2) implies that

$$\mathcal{E}(u,u) + \frac{c_4}{s^{2+\alpha}}||u||_1^2 \geq \frac{c_5}{s^2}||u||_2^2 \tag{3.1.3}$$

for any $u\in\mathcal{F}\cap L^1(K,\mu)$. By [**30**, Theorem 3.2], (3.1.3) turn out to be equivalent to the Nash inequality (A.1). Using Theorem A.2, we deduce that $\{T_t\}_{t>0}$ is ultracontractive and $||T_t||_{1\to\infty} \leq ct^{-\alpha/2}$. The existence of the heat kernel follows from Theorems A.2 and A.3.

If $(\mathcal{E},\mathcal{F},\mu)$ is recurrent, there exist $c>0$ and $\gamma>0$ such that $\mu(K_w)\geq c(r_w)^\gamma$ for any $w\in W_*$. Choose α so that $\gamma = (\alpha/2)/(1-(\alpha/2))$. Then $\alpha\in(0,2)$ and $\mu(K_w)\geq c_2 g(w)^\alpha$ for any $w\in W_*$. □

We also need the following two properties to establish a suitable framework for heat kernel estimate.

DEFINITION 3.1.7. Assume that $(\mathcal{E},\mathcal{F},\mu)$ satisfy the conditions (SSF) and (PI).
(1) $(\mathcal{E},\mathcal{F},\mu)$ is said to have the continuous heat kernel, (CHK) for short, if and if
(CHK) The heat kernel $p(t,x,y)$ associated with the Dirichlet form $(\mathcal{E},\mathcal{F})$ on $L^2(K,\mu)$ is jointly continuous, i.e. $p:(0,+\infty)\times K\times K\to[0,+\infty)$ is continuous.
(2) Let $(\Omega,\{X_t\}_{t\geq 0},\{P_x\}_{x\in K})$ be the diffusion process associated with the local regular Dirichlet form $(\mathcal{E},\mathcal{F})$ on $L^2(K,\mu)$. For any $A\subseteq K$, we define the hitting time of A, h_A, by $h_A = \inf\{t\geq 0|X_t\in A\}$. $(\mathcal{E},\mathcal{F},\mu)$ is said to have uniform positivity of hitting time, (UPH) for short, if and only if

(UPH) $\inf_{x\in B} E_x(h_A) > 0$ for all closed sets A and B with $A \cap B = \emptyset$.

In the subsequent sections, we will study heat kernels associated with a local regular Dirichlet form $(\mathcal{E}, \mathcal{F})$ on $L^2(K, \mu)$ which satisfy (SSF), (PI), (CHK) and (UPH). A similar set of assumptions on Dirichlet forms on self-similar sets has given in [**8**, Assumption 2.3]

In the recurrent case, (SSF) along with (PI) implies (CHK) and (UPH).

THEOREM 3.1.8. *Assume* (SSF) *and* (PI). *If* $(\mathcal{E}, \mathcal{F}, \mu)$ *is recurrent, then* (CHK) *and* (UPH) *are satisfied.*

LEMMA 3.1.9. *Assume* (SSF), (PI) *and that* $(\mathcal{E}, \mathcal{F}, \mu)$ *is recurrent. Then* $\mathcal{F} \subseteq C(K, d)$. *Let* U *be an open subset of* K. *Define* $\mathcal{F}_U = \{u | u \in \mathcal{F}, u|_{K\backslash U} = 0\}$ *and* $\mathcal{E}_U = \mathcal{E}|_{\mathcal{F}_U\times\mathcal{F}_U}$. *Also let* $\mu|_U$ *be the Borel regular measure on* U *defined by* $\mu|_U(A) = \mu(A)$ *for any Borel subset* A *of* U. *Then* $(\mathcal{E}_U, \mathcal{F}_U)$ *is a local regular Dirichlet form on* $L^2(U, \mu|_U)$. *The associated semigroup,* $\{T_t^U\}_{t>0}$, *on* $L^2(U, \mu|_U)$ *is ultracontractive and the associated heat kernel* $p_U : (0, +\infty) \times K \times K \to [0, +\infty)$ *is continuous.*

The heat kernel p_U itself is only defined on $(0, +\infty) \times U \times U$ by definition. However, we can extend $p_U(t, x, y)$ by letting $p_U(t, x, y) = 0$ if x or y belongs to $K\backslash U$.

PROOF. By Theorem 3.1.4, it follows that $||T_t||_{1\to\infty} < ct^{\alpha/2}$, where $\alpha \in (0, 2)$. Hence applying Theorem A.6, we obtain that $\mathcal{F} \subseteq C(K, d)$. Then $(\mathcal{E}_U, \mathcal{F}_U)$ is a local regular Dirichlet form on $L^2(U, \mu|_U)$ by [**15**, Theorem 4.4.3]. Starting from (3.1.3), we follow the same discussion as in the proof of Theorem 3.1.4 and obtain $||T_t^U||_{1\to\infty} \leq ct^{-\alpha/2}$. Hence Theorem A.6 shows that p_U is continuous. □

PROOF OF THEOREM 3.1.8. We already verify (CHK) in Lemma 3.1.9. Let A be a non-empty closed subset of K. For $x \in K$,

$$E_x(h_A) = \int_0^\infty \int_X p_Y(t, x, y)\mu(dy)dt, \tag{3.1.4}$$

where $Y = A^c$. Since $A \neq \emptyset$, $\mathcal{E}_Y(u, u) = 0$ if and only if $u = 0$. Therefore, if λ_1 is the smallest eigenvalue of the non-negative self-adjoint operator associated with the Dirichlet form $(\mathcal{E}_Y, \mathcal{F}_Y)$ on $L^2(Y, \mu|_Y)$, $-\Delta_Y$, then $\lambda_1 > 0$. By (A.2), there exists $c_1 > 0$ such that

$$p_Y(t, x, y) \leq c_1 e^{-\lambda t}$$

for any $x, y \in K$ and any $t \geq 1$. For $t \in (0, 1]$, By Lemma 3.1.9, there exists $c_2 > 0$ such that $p(t, x, y) \leq c_2 t^{-\alpha/2}$ for any $x, y \in K$ and $t \in (0, 1]$. Therefore, define

$$F(t) = \begin{cases} c_2 t^{-\alpha/2} & \text{if } t \in (0, 1], \\ c_1 e^{-\lambda_1 t} & \text{if } t > 1. \end{cases}$$

Then $p_Y(t, x, y) \leq F(t)$ and $\int_0^1 \int_X F(t)\mu(dy)dt < +\infty$. Note that $p_Y(t, x, y)$ is continuous. By the Lebesgue dominated convergence theorem, (3.1.4) implies that $E_x(h_A)$ is continuous with respect to $x \in K$. Assume that B is a closed subset of K and $A \cap B = \emptyset$. Since the process is a diffusion process, $P_x(h_A = 0) > 0$ for any $x \in B$. Hence $E_x(h_A) > 0$ for any $x \in B$. Therefore, $\inf_{x\in B} E_x(h_A) = \min_{x\in B} E_x(h_A) > 0$. Thus we obtain (UPH). □

PROPOSITION 3.1.10. *Assume* (SSF), (PI) *and* (CHK). *If* $(\mathcal{E}.\mathcal{F})$ *is conservative, i.e.* $1 \in \mathcal{F}$ *and* $\mathcal{E}(1,1) = 0$, *then the heat kernel* $p(t,x,y)$ *is positive, i.e.* $p(t,x,y) > 0$ *for any* $(t,x,y) \in (0,+\infty) \times K \times K$.

PROOF. By [**28**, Theorem 1.6.2], K is arcwise connected. Hence the desired result follows from Theorem A.4. □

3.2. Heat kernel estimate

In this section, we will give our main result on heat kernels associated with self-similar Dirichlet forms on self-similar sets. Let $\mathcal{L} = (K, S, \{F_i\}_{i\in S})$ be a self-similar structure. Hereafter we will always assume that $K \neq \overline{V}_0$, that K is connected and that $(K, S, \{F_i\}_{i\in S})$ is rationally ramified with a relation set $\mathcal{R}$. Moreover, μ is an elliptic probability measure on K and $(\mathcal{E},\mathcal{F})$ is a local regular Dirichlet form on $L^2(K,\mu)$.

DEFINITION 3.2.1. Assume that $(\mathcal{E},\mathcal{F},\mu)$ satisfy (SSF). The resistance scaling ratio $(r_i)_{i\in S}$ of $(\mathcal{E},\mathcal{F})$ is said to be arithmetic on $\mathcal{R}_1$-relations if and only if $\log r_w / \log r_v \in \mathbb{Q}$ for any $(\{w\},\{v\},\varphi,x,y) \in \mathcal{R}_1$.

For the lower off-diagonal estimate of heat kernels, we need a "geodesic" between a pair of points.

DEFINITION 3.2.2. Let (X,d) be a metric space. For $x,y \in X$, a curve $\gamma : [0,d(x,y)] \to X$ is called a geodesic between x and y if and only if $\gamma(0) = x, \gamma(1) = y$ and $d(\gamma(t),\gamma(s)) = |t-s|$ for any $t,s \in [0,d(x,y)]$. We call $(x,y) \in X^2$ a geodesics pair for (X,d) if and only if there exists a geodesic between x and y. The distance d is called a geodesic distance if and only if every pair $(x,y) \in X^2$ is a geodesic pair.

THEOREM 3.2.3. *Assume that* $(\mathcal{E},\mathcal{F})$ *is conservative and that* $(\mathcal{E},\mathcal{F},\mu)$ *satisfy* (SSF), (PI), (CHK) *and* (UPH). *Let* $\mathcal{S}_*$ *be the scale induced by the gauge function* $g(w) = \sqrt{r_w\mu(K_w)}$. *Suppose either that*
(I) $(\mathcal{E},\mathcal{F})$ *is recurrent*
or that
(II) μ *is a self-similar measure on* K *and the resistance scaling ratio* $(r_i)_{i\in S}$ *is arithmetic on* $\mathcal{R}_1$*-relations.*
Then, the following four conditions (a) - (d) *are equivalent.*
(a) μ *is volume doubling with respect to the scale* $\mathcal{S}_*$.
(b) *There exists a qdistance* d *on* K *adapted to* $\mathcal{S}_*$ *such that* μ *is volume doubling with respect to the qdistance* d.
(c) *There exist* $c > 0$ *such that*

(DUHK')
$$p(t,x,x) \le \frac{c}{\mu(U_{\sqrt{t}}(x))}$$

for any $t \in (0,1]$ *and any* $x \in K$.
(d) *There exist a qdistance* d *on* K *which is adapted to* $\mathcal{S}_*$ *and* $c > 0$ *such that*

(DUHK)
$$p(t,x,x) \le \frac{c}{\mu(B_{\sqrt{t}}(x,d))}$$

for any $t \in (0,1]$ *and any* $x \in K$.

Moreover, suppose that any of the above conditions holds. Let d be a qdistance adapted to $\mathcal{S}_*$. If d^α is a distance on K, then $\alpha < 2$ and there exist positive constants c_1, c_2 and c_3 such that, for any $t \in (0,1]$ and any $x, y \in K$,

$$\text{(DLHK)} \qquad \frac{c_1}{\mu(B_{\sqrt{t}}(x,d))} \leq p(t,x,x)$$

and

$$\text{(UHK)} \qquad p(t,x,y) \leq \frac{c_2}{\mu(B_{\sqrt{t}}(x,d))} \exp\Big(-c_3\Big(\frac{d(x,y)^2}{t}\Big)^{\frac{1}{\beta-1}}\Big),$$

where $\beta = 2/\alpha$. Also in the recurrent case, there exist positive constants c_4 and c_5 such that

$$\text{(LHK)} \qquad \frac{c_4}{\mu(B_{\sqrt{t}}(x,d))} \exp\Big(-c_5\Big(\frac{d(x,y)^2}{t}\Big)^{\frac{1}{\beta-1}}\Big) \leq p(t,x,y)$$

for any $t \in (0,1]$ and any geodesic pair $(x,y) \in K^2$ for (K, d^α).

REMARK. At a glance, it seems that the inequalities (DUHK), (DLHK) and (UHK) may depend on the choice of a qdistance d. Using $U_s(x)$ and $\delta^{(1)}(x,y)$, however, we may rewrite those inequalities. Namely, if $\delta^{(1)}(x,y)^\alpha$ is equivalent to a distance on K, then

$$\frac{\gamma_1}{\mu(U_{\sqrt{t}}(x))} \leq p(t,x,x) \leq \frac{\gamma_2}{\mu(U_{\sqrt{t}}(x))}$$

and

$$p(t,x,y) \leq \frac{\gamma_2}{\mu(U_{\sqrt{t}}(x))} \exp\Big(-\gamma_3\Big(\frac{\delta^{(1)}(x,y)^2}{t}\Big)^{\frac{1}{\beta-1}}\Big),$$

where $\beta = 2/\alpha$. Note that γ_1 and γ_2 are independent of α. The constant γ_3 is the only place where the value of α may be involved.

We will give a proof of Theorem 3.2.3 in Section 3.5.

There are two classes of self-similar sets, p. c. f self-similar sets and Sierpinski carpets, where a local regular Dirichlet form with (SSF), (PI), (CHK) and (UPH) has been constructed. We will apply the above theorem to those classes in the next two sections.

3.3. P. c. f. self-similar sets

In this section, we will consider post critically finite self-similar structures. In this case, one can easily determine when the assumptions of Theorem 3.2.3 hold. Throughout this section, $\mathcal{L} = (K, S, \{F_i\}_{i\in S})$ is a post critically finite self-similar structure whose relation set is $\{(\{w(i)\}, \{v(i)\}, \varphi_i, x(i), y(i)) | i = 1, \ldots, m\}$, where $w(i), v(i), x(i), y(i) \in W_\#$ and $\varphi_i(w(i)) = v(i)$.

There is an established way of constructing self-similar Dirichlet forms on a post critically finite self-similar sets in [**28**]. It starts from a harmonic structure $(D, \mathbf{r})$, where D is a "Laplacian" on V_0, which is a finite set for a p. c. f. self-similar set, and $\mathbf{r} = (r_i)_{i\in S} \in (0,\infty)^S$. From $(D, \mathbf{r})$, we obtain a quadratic form $(\mathcal{E}, \mathcal{F})$ which satisfies $u \circ F_i \in \mathcal{F}$ for any $i \in S$ and

$$\mathcal{E}(u,u) = \sum_{i\in S} \frac{1}{r_i} \mathcal{E}(u \circ F_i, u \circ F_i)$$

for any $u \in \mathcal{F}$. See [**28**] for details.

We assume that μ is an elliptic probability measure on K for the rest of this section.

PROPOSITION 3.3.1. *Assume either that* $(D, \mathbf{r})$ *is recurrent, i.e.* $\mathbf{r} \in (0,1)^S$, *or that* μ *is a self-similar measure with weight* $(\mu_i)_{i\in S}$ *which satisfies* $r_i\mu_i < 1$ *for any* $i \in S$. *Then* $(\mathcal{E}, \mathcal{F})$ *is an local regular Dirichlet form on* $L^2(K, \mu)$ *which satisfies* (SSF), (PI), (CHK) *and* (UPH).

PROOF. If $(D, \mathbf{r})$ is recurrent, then the conditions (RFA1), (RFA2) and (RFA3) are immediately verified. Hence the statement follows by Theorem B.3. Next assume that μ is a self-similar measure with weight $(\mu_i)_{i\in S}$ which satisfies $r_i\mu_i < 1$ for any $i \in S$. Then we have (SSF) by the method of construction of $\mathcal{F}$. See [**28**, Sections 3.1 and 3.2] for details. Also by [**28**, Theorem 3.4.6], $(\mathcal{E}, \mathcal{F})$ is a local regular Dirichlet form on $L^2(K, \mu)$ and the associated non-negative self-adjoint operator H has compact resolvent. Also the kernel of H is equal to constants. Therefore, by the remark after Definition 3.1.3, we obtain (PHI). By [**28**, Prposition 5.1.2], we also have (CHK). Finally, we show (UPH). Let A and B be closed subsets of K with $A \cap B = \emptyset$. Set $A_m = K(W(W_m, A))$. Then $A_m \cap B = \emptyset$ for sufficiently large m. Since $A \subseteq A_m$, we have $E_x(h_{A_m}) \leq E_x(h_A)$. Therefore, we may replace A by A_m to show (UPH). In other word, we may regard A as $\cup_{w\in\Gamma}K_w$ for some finite subset Γ of W_*. In such a case, ∂A is a finite subset of V_* and $h_A = h_{\partial A}$ for any path starting from B. By [**28**, Section A.2], the heat kernel $p_{\partial A}(t, x, y)$ corresponding to the Dirichlet form $(\mathcal{E}, \mathcal{F}_{\partial A})$ on $L^2(K, \mu)$ is jointly continuous on $(0, \infty) \times K^2$. Also, we have

$$E_x(h_A) = \int_{K\backslash A}\int_0^\infty p_{\partial A}(t, x, y)dt\mu(dy)$$

for any $x \in K\backslash A$. Define

$$F(x) = \int_{K\backslash A}\int_0^\infty p_{\partial A}(t, x, y)dt\mu(dy).$$

By definition, $0 \leq F(x) \leq E_x(h_A)$ for any $x \in B$. By [**28**, Theorem A.2.1], the nonnegative self-adjoint operator $-\Delta_{\partial A}$ associated with $(\mathcal{E}, \mathcal{F}_{\partial A})$ on $L^2(K, \mu)$ has compact resolvent. Let λ_* be the smallest eigenvalue of $-\Delta_{\partial A}$. If $\mathcal{E}(u, u) = 0$, then u is constant and $u|_{\partial A} \equiv 0$. This implies that $\lambda_* > 0$. Hence there exists $C > 0$ such that

$$p_{\partial A}(t, x, y) \leq Ce^{-\lambda_* t}$$

for any $(t, x, y) \in [1, \infty) \times K^2$. Hence $F(x)$ is continuous on $K\backslash A$ by the Lebesgue dominated convergence theorem. Moreover by [**28**, Theorem A.2.19], we have $p_{\partial A}(t, x, x) > 0$ for any $x \in K\backslash A$. Hence $F(x) > 0$ for any $x \in K\backslash A$. Since B is compact, we deduce that $0 < \inf_{x\in B} F(x) \leq \inf_{x\in B} E_x(h_A)$. Thus we obtain (UPH). $\square$

From now on, we confine ourselves to the second case in the above proposition, namely, μ is a self-similar measure with weight $(\mu_i)_{i\in S}$ which satisfies $r_i\mu_i < 1$ for any $i \in S$. Note that if $(D, \mathbf{r})$ is recurrent, then the assumption (I) of Theorem 3.2.3 is satisfied. If not, the resistance scaling ratio $\mathbf{r}$ should be arithmetic on $\mathcal{R}_1$-relations in order to satisfy the assumption (II) of Theorem 3.2. Note that every relation is an $\mathcal{R}_1$-relation for p. c. f. self-similar structure.

PROPOSITION 3.3.2. *The assumption (II) of Theorem 3.2.3 holds if and only if* $\log r_{w(i)}/\log r_{v(i)} \in \mathbb{Q}$ *for any* $i = 1, \ldots, m$.

PROOF. This is immediate by Definition 3.2.1. □

We have the following simple condition which is equivalent to the statement (a) of Theorem 3.2.3

PROPOSITION 3.3.3. *Let* $\mathcal{S}_*$ *be the scale induces by the gauge function* $g(w) = \sqrt{r_w \mu_w}$. *Then* μ *has the volume doubling property with respect to* $\mathcal{S}_*$ *if and only if*

$$\frac{\log r_{w(i)}}{\log \mu_{w(i)}} = \frac{\log r_{v(i)}}{\log \mu_{v(i)}}$$

for any $i = 1, \ldots, m$.

PROOF. Corollary 1.6.13 suffices to show the desired statement. □

If μ has the volume doubling property with respect to $\mathcal{S}_*$, we can apply Theorem 3.2.3 and obtain heat kernel estimates. As is seen in the last section, if

$$\max\{\alpha | D_{\gamma^\alpha} \text{ is a distance on } K\} \tag{3.3.1}$$

exists, then it plays an important role in off-diagonal heat kernel estimates like (UHK) and (LHK). Next we study how to calculate the value of maximum in (3.3.1).

DEFINITION 3.3.4. (1) Define

$$\mathcal{CH}_m(x,y) = \{(w(j))_{j=1,\ldots,k} | (w(j))_{j=1,\ldots,k} \in \mathcal{CH}(x,y), \\ w(j) \in W_m \text{ for any } j = 1, \ldots, k\},$$

for $x, y \in K$ and $m \geq 0$. We regard $\mathcal{CH}_1(x,y)$ as a subset of $W_\#$ by identifying $(w(j))_{j=1,\ldots,k} \in \mathcal{CH}_1(x,y)$ with $w(1)w(2)\ldots w(k) \in W_\#$.
(2) $(\mathcal{A}, \tau)$ is called a recursive system of paths if $\mathcal{A}$ is a non-empty finite subset of

$$\bigcup_{p,q \in V_0 : p \neq q} \{(w,p,q), w \in \mathcal{CH}_1(p,q)\},$$

and $\tau : \mathcal{A} \to \cup_{n \geq 1} \mathcal{A}^n$ satisfies the following condition: $\tau((w,p,q)) \in \mathcal{A}^{|w|}$ for any $(w,p,q) \in \mathcal{A}$. If $\tau((w_1 \ldots w_k, p, q)) = ((w^{(j)}, p_j, q_j))_{j=1,\ldots,k}$, then $p = F_{w_1}(p_1), q = F_{w_k}(q_k)$ and $F_{w_j}(q_j) = F_{w_{j+1}}(p_{j+1})$ for any $j = 1, \ldots, k-1$.
(3) Let $(\mathcal{A}, \tau)$ be a recursive system of paths. $(\mathcal{A}, \tau)$ is called irreducible if and only if $\mathcal{B} = \mathcal{A}$ whenever $\mathcal{B} \subseteq \mathcal{A}$ and $(\mathcal{B}, \tau|_{\mathcal{B}})$ is a recursive recursive system of paths.
(4) Let $(\mathcal{A}, \tau)$ be recursive and let $\mathbf{a} = (a_j)_{j \in S} \in (0,1)^S$. Then the relation matrix $M = M_{\mathcal{A},\tau,\mathbf{a}}$ is a $\#\mathcal{A} \times \#\mathcal{A}$-matrix defined by

$$M_{\mathbf{w},\mathbf{w}'} = \sum_{j : \mathbf{w}^{(j)} = \mathbf{w}'} a_{w_j}$$

where $\mathbf{w} = (w_1 \ldots w_k, p, q)$ and $\tau(\mathbf{w}) = ((\mathbf{w}^{(1)}, \ldots, \mathbf{w}^{(k)}))$.

In some cases, the following results are useful in determining whether $D_{\mathbf{a}}$ is a distance or not. In fact, later in this section, we will make use of them to characterize the value (3.3.1) for an example.

PROPOSITION 3.3.5. *Let* $\mathbf{a} = (a_j)_{j \in S} \in (0,1)^S$.
(1) *If* $\sum_{j=1}^k a_{w_j} \geq 1$ *for any* $w_1 \ldots w_k \in \cup_{p,q \in V_0 : p \neq q} \mathcal{CH}_1(p,q)$, *then* $D_{\mathbf{a}}$ *is a distance on* K.
(2) *If there exists a recursive* $(\mathcal{A}, \tau)$ *such that the maximum eigenvalue of the relation matrix* $M_{\mathcal{A},\tau,\mathbf{a}}$ *is less than one, then* $D_{\mathbf{a}}$ *is not a distance.*

The following notations are convenient in proving the above proposition.

NOTATION. Let $\mathbf{w} = (w(j))_{j=1,\ldots,k} \in \mathcal{CH}(x, y)$.
(1) For $\mathbf{v} = (v(j))_{j=1,\ldots,l} \in \mathcal{CH}(y, z)$, we use $\mathbf{w} \vee \mathbf{v}$ to denote the chain between x and z defined by $((w(1), \ldots, w(k), v(1), \ldots, v(l))$. In the same manner, we define $\vee_{i=1}^n \mathbf{w}_i \in \mathcal{CH}(x_1, x_n)$ if $\mathbf{w}_i \in \mathcal{CH}(x_i, x_{i+1})$ $i = 1, \ldots, n-1$.
(2) For $v \in W_*$, define $v\mathbf{w} = (vw(j))_{j=1,\ldots,k}$, which is a chain between $F_v(x)$ and $F_v(y)$.
(3) For $\mathbf{a} = (a_j)_{j \in S} \in (0,1)^S$ and $\mathbf{w} = (w(j))_{j=1,\ldots,k} \in \mathcal{CH}$, define

$$\mathbf{a}_{\mathbf{w}} = \sum_{j=1}^{k} a_{w(j)}.$$

PROOF. (1) Assume that $p, q \in V_0$ and $p \neq q$. Let $(w(j))_{j=1,\ldots,n} \in \mathcal{CH}(p, q)$. We will show that

$$\sum_{j=1}^{n} a_{w(j)} \geq 1 \tag{3.3.2}$$

by using induction on n. If $n = 1$, then $w(1) = \emptyset$. Since $a_\emptyset = 1$, (3.3.2) holds. Also if $(w(j))_{j=1,\ldots,n} \in \mathcal{CH}_1(p, q)$, then (3.3.2) also holds by the assumption of the proposition. Otherwise, there exist $w \in W_\#$, $j_* \in \{1, \ldots, n\}$ and $i_1, \ldots, i_l \in S$ such that $w(j_* + k - 1) = wi_k$ for $k = 1, \ldots, l$ and $(i_k)_{k=1,\ldots,l} \in \mathcal{CH}_1(p', q')$ for some $p', q' \in V_0$ with $p' \neq q'$. Let $(w'(1), \ldots, w'(n-l+1)) = (w(1), \ldots, w(j_* - 1), w, w(j_* + l), \ldots, w(n))$. Then $(w'(1), \ldots, w'(n-l+1)) \in \mathcal{CH}(p, q)$. Using the assumption of the proposition and induction, we obtain

$$\sum_{j=1}^{n} a_{w(j)} = \sum_{j=1}^{j_*-1} a_{w(j)} + a_w \sum_{k=1}^{l} a_{i_k} + \sum_{j=j_*+l}^{n} a_{w(j)} \geq \sum_{m=1}^{n-l+1} a_{w'(m)} \geq 1.$$

Therefore, (3.3.2) holds for any element of $\mathcal{CH}(p, q)$. This immediately implies that $D_{\mathbf{a}}(p, q) \geq a_w$ for any $w \in W_*$ and any $p, q \in F_w(V_0)$ with $p \neq q$.

Next, define $K_m(x) = \cup_{w \in W_m : x \in K_w} K_w$. For any $x, y \in K$ with $x \neq y$, we may choose $m \geq 1$ such that $K_m(x) \cap K_m(y) = \emptyset$. Then for any $(w(j))_{j=1,\ldots,n} \in \mathcal{CH}(x, y)$, there exist j_*, l and $p, q \in V_m$ with $p \neq q$ such that $(w(j_*), \ldots, w(j_* + l - 1)) \in \mathcal{CH}(p, q)$. This shows that $D_{\mathbf{a}}(x, y) \geq \min_{p,q \in V_m : p \neq q} D_{\mathbf{a}}(p, q) > 0$. Thus $D_{\mathbf{a}}$ is a distance.

(2) Let $(\mathcal{A}, \tau)$ be a recursive system of paths. First define $\tau_m(w, p, q) \in \mathcal{CH}_m(p, q)$ for $(w, p, q) \in \mathcal{A}$ inductively as follows. Set $\tau_1(w, p, q) = w$ for any $(w, p, q) \in \mathcal{A}$. If $\tau(w, p, q) = ((w(j), p_j, q_j))_{j=1,\ldots,k}$, then we define $\tau_m(w, p, q) = \vee_{j=1}^k w_j \tau_{m-1}(w(j), p_j, q_j)$, where $w = w_1 \ldots w_k$.

Let $\mathbf{a} = (a_j)_{j \in S} \in (0,1)^S$. Then $\mathbf{a}_{\tau_m(w,p,q)} = (M^m e)_{(w,p,q)}$ for any $(w, p, q) \in \mathcal{A}$, where $M = M_{A,\tau,\mathbf{a}}$ and $e \in \ell(\mathcal{A})$ is the transpose of $(1, \ldots, 1)$. Assume that the maximum eigenvalue of M is less than one. It follows that the maximum eigenvalue of $M_{\mathcal{B},\tau|_{\mathcal{B}},\mathbf{a}}$ is less than one if $\mathcal{B} \subseteq \mathcal{A}$ and $(\mathcal{B}, \tau|_{\mathcal{B}})$ is a recursive system of paths. On the other hand, there exists an irreducible recursive system of paths $(\mathcal{B}, \tau')$ where $\mathcal{B} \subseteq \mathcal{A}$ and $\tau' = \tau|_{\mathcal{B}}$. Therefore, $\mathcal{A}$ may be assumed to be irreducible without loss of generality. Let λ be the maximum eigenvalue of M Then by the Perron-Frobenius theorem, $0 < \lambda < 1$ and we can choose a positive vector f as an associated eigenvector. Since $e \leq cf$ for some $c > 0$, $M^n e \leq c\lambda^n f$ as $n \to \infty$. Hence

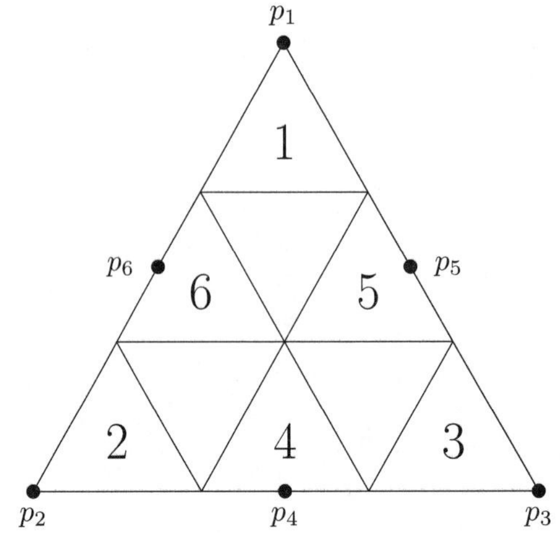

FIGURE 3.1. the modified Sierpinski gasket

$\lim_{m\to\infty} \mathbf{a}_{\tau_m(w,p,q)} = 0$. This implies that $D_{\mathbf{a}}(p,q) = 0$ if $(w,p,q) \in \mathcal{A}$. Therefore $D_{\mathbf{a}}$ is not a distance. □

Finally, we apply the above results to a particular example.

DEFINITION 3.3.6. Set $p_1 = e^{\sqrt{-1}\pi/6}, p_2 = 0, p_3 = 1, p_4 = (p_2 + p_3)/2, p_5 = (p_3 + p_1)/2$ and $p_6 = (p_1 + p_2)/2$. Define $F_i : \mathbb{C} \to \mathbb{C}$ by $F_i(z) = (z - p_i)/3 + p_i$ for $i = 1, \dots, 6$. Let K be the unique non-empty compact set that satisfies $K = \cup_{i\in S} F_i(K)$, where $S = \{1, \dots, 6\}$. K is called the modified Sierpinski gasket.

In the rest of this section, K is assumed to be the modified Sierpinski gasket and $\mathcal{L} = (K, S, \{F_i\}_{i\in S})$ is the associated self-similar structure defined above. Immediately by the above definition, we obtain the following.

PROPOSITION 3.3.7. *The relation set of $\mathcal{L}$ is*

$$\begin{aligned}\{(\{1\},\{2\},\varphi_{12},i,j)|(i,j) &= (6,1),(2,6),(4,5)\}\\ \cup\{(\{2\},\{3\},\varphi_{23},i,j)|(i,j) &= (4,2),(3,4),(5,6)\}\\ \cup\{(\{3\},\{1\},\varphi_{31},i,j)|(i,j) &= (1,5),(5,3),(6,4)\},\end{aligned}$$

where $\varphi_{kl}(k) = l$ for $(k,l) = (1,2),(2,3),(3,1)$. In particular, $\mathcal{L}$ is post critically finite, $\mathcal{P} = \{(1)^\infty, (2)^\infty, (3)^\infty)$ and $V_0 = \{p_1, p_2, p_3\}$.

PROPOSITION 3.3.8. *Let $D = \begin{pmatrix} -2 & 1 & 1 \\ 1 & -2 & 1 \\ 1 & 1 & -2 \end{pmatrix}$ and let $\mathbf{r} = (\frac{7}{15}, \dots \frac{7}{15})$.*
(1) *$(D, \mathbf{r})$ is a recurrent harmonic structure on $(K, S, \{F_i\}_{i\in S})$.*
(2) *Let μ be a self-similar measure on K with weight $(\mu_i)_{i\in S}$ and let $\mathcal{S}_*$ be a self-similar scale with weight $\{\gamma_i\}_{i\in S}$, where $\gamma_i = \sqrt{\mu_i r_i}$. Then μ has the volume doubling property with respect to $\mathcal{S}_*$ if and only if $\mu_1 = \mu_2 = \mu_3$.*

PROOF. (1) This can be shown by the Δ-Y transform. See [**28**] for details.
(2) Apply Proposition 3.3.3. □

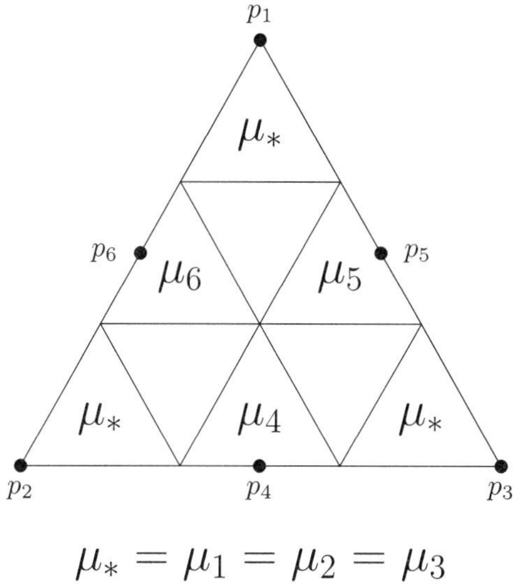

FIGURE 3.2. Self-similar volume doubling measures on the modified Sierpinski gasket

Hereafter, we fix $(D, \mathbf{r})$ and $\mathcal{S}_*$ as in the above proposition. Also μ is assumed to be a self-similar measure which satisfies $\mu_1 = \mu_2 = \mu_3$. See Figure 3.2. Note that we may assume that $\mu_4 \leq \mu_5 \leq \mu_6$ without loss of generality.

PROPOSITION 3.3.9. *Assume that $\mu_4 \leq \mu_5 \leq \mu_6$. Let α_* be the unique α which satisfies*

$$2(\gamma_1)^\alpha + (\gamma_4)^\alpha = 1. \tag{3.3.3}$$

Then

$$\alpha_* = \max\{\alpha | D_{\gamma^\alpha} \text{ is a distance on } K\},$$

where $\gamma^\alpha = ((\gamma_i)^\alpha)_{i=1,\dots,6}$.

Note that $\gamma_i = \sqrt{7\mu_i/15}$ for any i.

PROOF. Let $\mathbf{w} = (243, p_2, p_3)$. Note that $243 \in \mathcal{CH}_1(p_2, p_3)$. Set $\mathcal{A} = \{\mathbf{w}\}$ and define $\tau : \mathcal{A} \to \mathcal{A}^3$ by $\tau(\mathbf{w}) = (\mathbf{w}, \mathbf{w}, \mathbf{w})$. Then $(\mathcal{A}, \tau)$ is a recursive system of paths and $M_{\mathcal{A},\tau,\gamma^\alpha} = (2(\gamma_1)^\alpha + (\gamma_4)^\alpha)$. If $\alpha > \alpha_*$, the maximum eigenvalue of $M_{\mathcal{A},\tau,\gamma^\alpha}$ is less than one. Hence Proposition 3.3.5-(2) implies that D_{γ^α} is not a distance. On the other hand, for $\alpha = \alpha_*$, we may verify the assumption of Proposition 3.3.5-(1) and show that D_{γ^α} is a distance. □

THEOREM 3.3.10. *Assume that $\mu_4, \leq \mu_5 \leq \mu_6$. Let $(\mathcal{E}, \mathcal{F})$ be the Dirichlet form associated with $(D, \mathbf{r})$ on $L^2(K, \mu)$ and let $p(t, x, y)$ be the corresponding heat kernel.Also define $d = (D_{\gamma^{\alpha_*}})^{1/\alpha_*}$, where α_* is the unique solution of* (3.3.3).
(1) *Suppose that $\mu_4 < \mu_5$. Then,* (UHK) *and* (DLHK) *holds for any $x, y \in K$ and any $t \in (0, 1]$ with $\beta = \beta_*$. Moreover,* (LHK) *holds if the line segment $\overline{xy}$ is contained in K and is parallel to the real axis.*

(2) *If* $\mu_4 = \mu_5$, *then* (UHK) *and* (LHK) *holds for any* $x, y \in K$ *and any* $t \in (0, 1]$ *with* $\beta = \beta_*$.

PROOF. In both cases, Theorem 3.2.3 immediately implies (UHK). Assume that the line segment $\overline{xy}$ is contained in K and is parallel to the real axis. Then we see that (x, y) is a geodesic pair for $(K, D_{\gamma^{\alpha_*}})$. Hence by Theorem 3.2.3, we have (LHK) for such a pair. In the case (2), it follows that $D_{\gamma^{\alpha_*}}$ is equivalent to a geodesic distance. Hence (LHK) holds for any $x, y \in K$. □

3.4. Sierpinski carpets

In this section, we discuss another class of self-similar sets, the generalized Sierpinski carpets. The following definition is given by Barlow-Bass[**7**].

DEFINITION 3.4.1. Let $H_0 = [0, 1]^n$, where $n \in \mathbb{N}$, and let $l \in \mathbb{N}$ with $l \geq 2$. Set $\mathcal{Q} = \{\prod_{i=1}^n [(k_i - 1)/l, k_i/l] \,|\, (k_1, \ldots, k_n) \in \{1, \ldots, l\}^n\}$. For any $Q \in \mathcal{Q}$, define $F_Q : H_0 \to H_0$ by $F_Q(x) = x/l + a_Q$, where we choose a_Q so that $F_Q(H_0) = Q$. Let $S \subseteq \mathcal{Q}$ and let $\mathrm{GSC}(n, l, S)$ be the self-similar set with respect to $\{F_Q\}_{Q \in S}$, i.e. $\mathrm{GSC}(n, l, S)$ is the unique nonempty compact set satisfying $\mathrm{GSC}(n, l, S) = \cup_{Q \in S} F_Q(\mathrm{GSC}(n, l, S))$. Set $H_1(S) = \cup_{Q \in S} F_Q(H_0)$. $\mathrm{GSC}(n, l, S)$ is called a generalized Sierpinski carpet if and only if the following four conditions (GSC1), ..., (GSC4) are satisfied:
(GSC1) (Symmetry) $H_1(S)$ is preserved be all the isometries of the unit cube H_0.
(GSC2) (Connected) $H_1(S)$ is connected.
(GSC3) (Non-diagonality) For any $x \in H_1(S)$, there exists $r_0 > 0$ such that $\mathrm{int}(H_1(S) \cap B_r(x))$ is nonempty and connected for any $r \in (0, r_0)$, where $B_r(x) = \{y | y \in \mathbb{R}^n, |x - y| < r\}$.
(GSC4) (Border included) The line segment between 0 and $(1, 0, \ldots, 0)$ is contained in $H_1(S)$.

The Sierpinski carpet (Example 1.7.4) is equal to $\mathrm{GSC}(2, 3, S)$, where $S = \mathcal{Q} - \{[1/3, 2/3]^2\}$. Also $[0, 1]^n = \mathrm{GSC}(n, l, \mathcal{Q})$ for any $l \geq 2$.

In the rest of this section, we fix a generalized Sierpinski carpet $\mathrm{GSC}(n, l, S)$ and write $K = \mathrm{GSC}(n, l, S)$. Also $\mathcal{L}$ is the self-similar structure associated with K, i.e. $\mathcal{L} = (\mathrm{GSC}(n, l, S), S, \{F_Q\}_{Q \in S})$. Let ν be a self-similar measure with weight $(1/N, \ldots, 1/N)$, where $N = \#(S)$.

DEFINITION 3.4.2. For $k \in \{1, \ldots, n\}$ and $s \in [0, 1]$, define $S_{k,s} = \{Q | Q \in S, Q \cap \Phi_{k,s}\}$, where $\Phi_{k,s}$ is a hyperplane in $\mathbb{R}^n$ defined by

$$\Phi_{k,s} = \{(x_1, \ldots, x_n) | x_k = s\}.$$

Also let $\Psi_{k,l}$ be the parallel translation in k-direction by $1/l$; $\Psi_{k,l}(x_1, \ldots, x_n) = (y_1, \ldots, y_n)$, where $y_i = x_i$ if $i \neq k$ and $y_k = x_k + 1/l$. For $Q_1, Q_2 \in S$, Q_1 and Q_2 are called k-neighbors if and only if $\Psi_{k,l}(Q_1) = Q_2$ or $\Psi_{k,l}(Q_2) = Q_1$.

Let $\mathrm{rf_k} : \mathrm{R^n} \to \mathrm{R^n}$ be the reflection in the hyperplane $\Phi_{k,1/2}$. The symmetry condition (GSC1) ensures that $\mathrm{rf_k(Q) \in S_{k,1}}$ for any $Q \in S_{k,1}$. In this sense, we regard $\mathrm{rf_k}$ as a map from $S_{k,0}$ to $S_{k,1}$. Note that $\mathrm{rf_k}$ is a bijection between $S_{k,0}$ and $S_{k,1}$.

PROPOSITION 3.4.3. *The self-similar structure $\mathcal{L}$ associated with a generalized Sierpinski carpet is rationally ramified with a relation set*

$$\mathcal{R}_* = \{(S_{k,0}, S_{k,1}, \mathrm{rf_k}, \mathrm{Q_1}, \mathrm{Q_2}) \,|\, \mathrm{k} \in \{1, \ldots, \mathrm{n}\}, \\ Q_1, Q_2 \in S \text{ and they are } k\text{-neighbors.}\}$$

Combining the above proposition with Theorem 1.6.1, we obtain the following fact.

PROPOSITION 3.4.4. *A self-similar scale $\mathcal{S}(\mathbf{a})$ is locally finite with respect to $\mathcal{L}$. if and only if $a_{\mathrm{rf_k(Q)}} = a_Q$ for any $k = 1, \ldots, n$ and any $Q \in S_{k,0}$.*

In the series of papers [**2, 3, 4, 5, 6, 7**], Barlow and Bass have constructed a diffusion process on a generalized Sierpinski carpet and studied it extensively. For example, they have obtained elliptic and parabolic Harnack inequalities, Poincaré inequality and sub-Gaussian heat kernel estimate. Unfortunately, the Dirichlet form on $L^2(K, \nu)$ associated with their diffusion process is not necessarily self-similar. On the other hand, in [**34**], Kusuoka and Zhou have given a prescription of construction a self-similar Dirichlet form on a generalized Sierpinski carpet.

Combining the methods and results in [**7**] and [**34**] as in [**22**], we obtain a local regular Dirichlet form $(\mathcal{E}, \mathcal{F})$ on $L^2(K, \nu)$ which has the self-similarity in the following sense: for any $u \in \mathcal{F}$ and any $Q \in S$, $u \circ F_Q \in \mathcal{F}$ and there exists $r > 0$ such that

$$\mathcal{E}(u, u) = \frac{1}{r} \sum_{Q \in S} \mathcal{E}(u \circ F_Q, u \circ F_Q)$$

for any $u \in Q$. In fact, from Kusuoka-Zhou's method, we have (SSF). Moreover, the corresponding diffusion process enjoys the same inequalities and estimates as the original one studied by Barlow and Bass. See [**7**, Remark 5.11] and the discussion after it. In particular, the associated heat kernel satisfies UHK and LHK for any $x, y \in K$, where $\beta > 2$, $\mu = \nu$ and a distance d is the Euclidean distance. Note that $\nu(B_r(x, d)) = cr^n$ for any $r > 0$.

Barlow-Kumagai have studied a time change of this process in [**8**]. Let μ be a self-similar measure on $\mathcal{L}$ with weight $(\mu_i)_{i \in S}$. Define

$$\mathcal{F}_\mu = \{u | u \in L^2(K, \mu), \text{there exists } f \in \mathcal{F}_e \text{ such that } u = f \text{ for } \mu\text{-a.e. } x \in K\}$$

and set $\mathcal{E}_\mu(u, u) = \mathcal{E}(\widetilde{H}_A f, \widetilde{H}_A f)$ for $u \in \mathcal{F}_\mu$, where $f \in \mathcal{F}_e$ and $u = f$ for μ-a.e. $x \in K$, A is the quasi support of μ and $(\widetilde{H}_A u)(x) = E_x(u(X_{h_A}))$. (See [**15**, Section 6.2] for details on time changes of a diffusion process associated with a Dirichlet form in general. Also see [**8**, p. 9].) In [**8**], they have shown that if $\mu_Q r < 1$ for any $Q \in S$, then $(\mathcal{E}_\mu, \mathcal{F}_\mu)$ is a local regular Dirichlet form on $L^2(K, \mu)$ and the associated diffusion process is a time change of the diffusion associated with $(\mathcal{E}, \mathcal{F})$. By their discussion, we can verify (SSF), (PI), (CHK) and (UPH).

Here after, we fix a self-similar measure μ with weight $(\mu_Q)_{Q \in S}$ and assume that $\mu_Q r < 1$ for any $Q \in S$. The following lemma is immediate by Theorems 1.3.5, 1.6.6 and Proposition 3.4.4.

THEOREM 3.4.5. *Define $\mathcal{S}_* = \mathcal{S}(\gamma)$, where $\gamma_Q = \sqrt{\mu_Q r}$ for any $Q \in S$ and $\gamma = (\gamma_Q)_{Q \in S}$. Then μ has the volume doubling property with respect to $\mathcal{S}_*$ if and only if $\mu_Q = \mu_{\mathrm{rf_k(Q)}}$ for any $k = 1, \ldots, n$ and any $Q \in S_{k,0}$.*

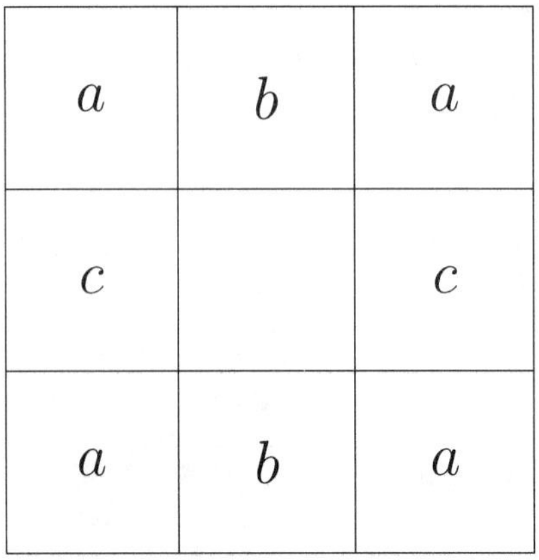

$$a = \mu_1 = \mu_3 = \mu_5 = \mu_7$$
$$b = \mu_2 = \mu_6$$
$$c = \mu_4 = \mu_8$$

FIGURE 3.3. Self-similar volume doubling measures on the Sierpinski carpet

This theorem shows when the condition (a) of Theorem 3.2.3 holds. Consequently, the claims of Theorem 3.2.3 follows if $\mu_Q = \mu_{\mathrm{rf_k(Q)}}$ for any $k = 1, \ldots, n$ and any $Q \in S_{k,0}$. In particular, if D_{γ^α} is a distance, set $\beta = 2/\alpha$ and $d(x, y) = D_{\gamma^\alpha}(x, y)^{1/\alpha}$. Then

$$\frac{c_1}{\mu(B_{\sqrt{t}}(x, d))} \leq p(t, x, x) \leq \frac{c_2}{\mu(B_{\sqrt{t}}(x, d))}$$

and

$$p(t, x, y) \leq \frac{c_3}{\mu(B_{\sqrt{t}}(x, d))} \exp\Big(- c_4\Big(\frac{d(x, y)^2}{t}\Big)^{\frac{1}{\beta - 1}}\Big)$$

for any $t \in (0, 1]$ and any $x, y \in K$, where $p(t, x, y)$ is the heat kernel associated with the Dirichlet form $(\mathcal{E}_\mu, \mathcal{F}_\mu)$ on $L^2(K, \mu)$. Moreover, we have the elliptic Harnack inequality by [**7**]. (Note that harmonic functions associated with $(\mathcal{E}_\mu, \mathcal{F}_\mu)$ on $L^2(K, \mu)$ are the same as those associated with $(\mathcal{E}, \mathcal{F})$ on $L^2(K, \nu)$.) Also we have the exit time estimate (E) by Lemma 3.5.13. Using the arguments in [**17**], we have the near diagonal lower estimate (3.5.8). Hence, if D_{γ^α} is equivalent to a geodesic distance, then the classical arguments in [**1, 8, 18, 30**] imply the lower off-diagonal Li-Yau estimate (LHK).

Finally we present two examples.

EXAMPLE 3.4.6 (the Sierpinski carpet). Let $\mathcal{L} = (K, S, \{F_i\}_{i \in S})$ be the self-similar structure associated with the Sierpinski carpet appearing in Examples 1.5.12 and 1.7.4. By [**2, 3**] and [**34**], the resistance scaling ratio r is less than one and hence we are in the recurrent case. By Theorem 3.4.5, the condition (a) of Theorem 3.2.3 follows if and only if $\mu_1 = \mu_3 = \mu_5 = \mu_7$, $\mu_2 = \mu_6$ and $\mu_4 = \mu_8$. See Figure 3.3. Furthermore, if $\mu_2 = \mu_4$ and $\mu_1 \leq \mu_4$ as well, then

$$(0, \alpha_*] = \{\alpha | D_{\gamma^\alpha} \text{ is a distance}\},$$

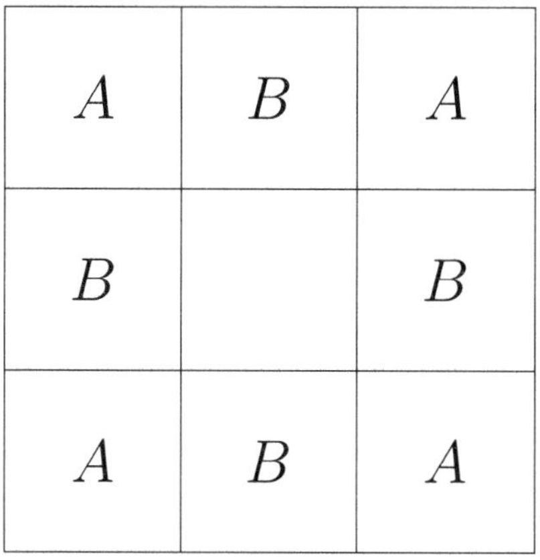

$$A = (\mu_1 r)^{\alpha_*/2}$$
$$B = (\mu_2 r)^{\alpha_*/2}$$
$$A \leq B$$
$$2A + B = 1$$

FIGURE 3.4. Geodesic distances on the Sierpinski carpet

where α_* is given by $2(\mu_1 r)^{\alpha_*/2} + (\mu_2 r)^{\alpha_*/2} = 1$, and $D_{\gamma^{\alpha_*}}$ is equivalent to a geodesic distance. See Figure 3.4. Details on the construction of geodesic distances on the Sierpinski carpet can be found in [**32**]. In this case, we have the upper and lower off-diagonal Li-Yau estimates (UHK) and (LHK):

$$\frac{c_1}{\mu(B_{\sqrt{t}}(x,d))} \exp\left(- c_2 \Big(\frac{d(x,y)^2}{t} \Big)^{1/(\beta_* - 1)} \right) \leq p(t,x,y)$$
$$\leq \frac{c_3}{\mu(B_{\sqrt{t}}(x,d))} \exp\left(- c_4 \Big(\frac{d(x,y)^2}{t} \Big)^{1/(\beta_* - 1)} \right)$$

for any $x, y \in K$ and any $t \in (0,1]$, where $d(x,y) = (D_{\gamma^{\alpha_*}})^{1/\alpha_*}$ and $\beta_* = 2/\alpha_*$.

EXAMPLE 3.4.7 (Cubes). Let $l = 3$ and let $S = \mathcal{Q}$. Then $K = [0,1]^n$. In this case, ν is the restriction of the Lebesgue measure,

$$\mathcal{F} = H_1(K) = \{f | f : K \to \mathbb{R}, \text{all the partial derivatives of } f$$
$$\text{in the sense of distribution belong to } L^2(K,\nu)\}$$

and

$$\mathcal{E}(u,v) = \sum_{k=1}^{n} \int_K \frac{\partial u}{\partial x_k} \frac{\partial v}{\partial x_k} d\nu,$$

where $\partial u / \partial x_i$ is the derivative in the sense of distribution. The diffusion process associated with the Dirichlet form $(\mathcal{E}, \mathcal{F})$ on $L^2(K,\nu)$ is the reflected Brownian motion. In this case, for any $u, v \in \mathcal{F}$,

$$\mathcal{E}(u,v) = 3^{2-n} \sum_{Q \in S} \mathcal{E}(u \circ F_Q, v \circ F_Q).$$

Hence $r = 3^{n-2}$. Hence we are not in the recurrent case unless $n = 1$. If $\mu_Q < 3^{2-n}$ for any $Q \in S$, then we have a local regular Dirichlet form $(\mathcal{E}_\mu, \mathcal{F}_\mu)$ on $L^2(K, \mu)$, where μ is the self-similar measure with weight $(\mu_Q)_{Q \in S}$. The corresponding diffusion process is the time change of the reflected Brownian motion on n-dimensional cube $[0,1]^n$. In particular, if $n = 2$, then $r = 1$ and $(\mathcal{E}_\mu, \mathcal{F}_\mu)$ is a local regular Dirichlet form on $L^2(K, \mu)$ for any self-similar measure μ. Applying Theorem 3.4.5, we obtain Theorem 0.2.5.

3.5. Proof of Theorem 3.2.3

As in Section 3.2, $(K, S, \{F_i\}_{i \in S})$ is a rationally finite self-similar structure and $(\mathcal{E}, \mathcal{F})$ is a local regular Dirichlet form on $L^2(K, \mu)$ which is conservative and satisfies (SSF), (PI), (CHK) and (UPH). Also $\mathcal{S}_*$ is the scale induced by the gauge function $g(w) = \sqrt{r_w \mu(K_w)}$. We write $\mathcal{S}_* = \{\Lambda_s\}_{0<s\leq 1}$.

First note that Theorem 2.3.17 implies the following equivalence.

LEMMA 3.5.1. (a) *is equivalent to* (b).

DEFINITION 3.5.2. Let U be a nonempty open subset of K. Define $\mathcal{D}_U = \{u | u \in \mathcal{F} \cap C(K), u|_{K \backslash U} \equiv 0\}$ and

$$\lambda_*(U) = \inf_{u \in \mathcal{D}_U} \frac{\mathcal{E}(u,u)}{||u||_2^2}.$$

Also define $\mathcal{F}_U$ by the closure of $\mathcal{D}_U$ with respect to the inner product $\mathcal{E}_*(u,v) = \mathcal{E}(u,v) + \int_K uv d\mu$.

PROPOSITION 3.5.3. *Let U be a nonempty open subset of K. If $\mathcal{E}_U = \mathcal{E}_{\mathcal{F}_U \times \mathcal{F}_U}$, then $(\mathcal{E}_U, \mathcal{F}_U)$ is a local regular Dirichlet form on $L^2(K, \mu)$ (or $L^2(U, \mu|_U)$). If $-\Delta_U$ is the self-adjoint operator on $L^2(K, \mu)$ associated with $(\mathcal{E}_U, \mathcal{F}_U)$, then $-\Delta_U$ has compact resolvent and $\lambda_*(U)$ is the minimal eigenvalue of $-\Delta_U$. Also if $p_U(t,x,y)$ is the heat kernel associated with the Dirichlet form $(\mathcal{E}_U, \mathcal{F}_U)$, then. for any $t > 0$,*

$$0 \leq p_U(t,x,y) \leq p(t,x,y)$$

for $\mu \times \mu$-a.e. $(x,y) \in K^2$.

LEMMA 3.5.4. *There exists $c > 0$ such that, for any $w \in W_*$,*

$$\lambda_*(B_w) \leq \frac{c}{r_w \mu(K_w)},$$

where $B_w = K_w \backslash F_w(\overline{V}_0)$.

PROOF. Choose $v \in W_*$ so that $K_v \subseteq K \backslash \overline{V}_0$. Since $(\mathcal{E}, \mathcal{F})$ is a regular Dirichlet form, there exists $\varphi \in C(K) \cap \mathcal{F}$ such that $\mathrm{supp}(\varphi) \subseteq K \backslash \overline{V}_0$ and $\varphi(x) \geq 1$ for any $x \in K_v$. Define φ_w by

$$\varphi_w(x) = \begin{cases} \varphi((F_w)^{-1}(x)) & \text{if } x \in K_w, \\ 0 & \text{otherwise.} \end{cases}$$

Then by (SSF), $\varphi_w \in \mathcal{F}_{B_w}$ and $\mathcal{E}(\varphi_w, \varphi_w) = (r_w)^{-1}\mathcal{E}(\varphi, \varphi)$. Since μ is elliptic, $||\varphi_w||_2^2 \geq \mu(K_{wv}) \geq c'\mu(K_w)$, where c' is independent of w. Therefore,

$$\lambda_*(B_w) \leq \frac{\mathcal{E}(\varphi_w, \varphi_w)}{||\varphi_w||_2^2} \leq \frac{\mathcal{E}(\varphi, \varphi)}{c' r_w \mu(k_w)}.$$

□

LEMMA 3.5.5. (d) *implies* (b).

PROOF. Choose $\alpha > 0$ so that d^α is a distance. Let $D(x,y) = d(x,y)^\alpha$ and let $\beta = 2/\alpha$. Then by (d),

$$p(t,x,x) \leq \frac{c_1}{\mu(B_{t^{1/\beta}}(x,D))}.$$

Since d is adapted to the scale $\mathcal{S}_*$, there exists $c_2 > 0$ such that $U_{c_2 s}(x) \subseteq B_s(x,d)$ for any $x \in K$ and any $s \in (0,1]$. Hence for any $r > 0$, $U_{c_2 s}(x) \subseteq B_s(x,d) = B_r(x,D)$, where $s = r^{1/\alpha}$. Let $w \in \Lambda_{c_2 s,x}$. Then by Lemma 3.5.4,

$$\lambda_*(B(r,D)) \leq \lambda_*(B_w) \leq \frac{c_3}{r_w \mu(K_w)} \leq c_4 r^{-\beta}.$$

Using Theorem C.3, we have the volume doubling property of μ with respect to the distance D. This immediately implies (b). □

LEMMA 3.5.6. *If there exist positive constants c_1 and c_2 such that*

$$p(t,x,x) \leq \frac{c_1}{\mu(U_{c_2\sqrt{t}}(x))}$$

for any $x \in X$ and any $t \in (0,1]$, then μ has the volume doubling property with respect to $\mathcal{S}_*$. In particular, (c) *implies* (a).

REMARK. In the following proof, we don't need the assumption (I) neither (II).

PROOF. Let $s = c_2\sqrt{t}$ and let $x \in K$. If $w \in \Lambda_{s,x}$, then $U_s(y) = U_s(w) \subseteq U_s(x)$ for any $y \in B_w$, where $U_s(w) = K(W(\Lambda_s, K_w))$. By (c) and Proposition 3.5.3,

$$p_{B_w}(t,y,y) \leq p(t,y,y) \leq \frac{c_1}{\mu(U_s(y))}.$$

Integrating this over B_w, we see that

$$e^{-\lambda_*(B_w)t} \leq \int_{B_w} p_{B_w}(t,y,y)\mu(dy) \leq \frac{c_1\mu(K_w)}{\mu(U_s(y))}.$$

By Lemma 3.5.4, it follows that $c_* \leq e^{-\lambda_*(B_w)t}$, where c_* is independent of x, t and $w \in \Lambda_{s,x}$. Hence,

$$c_*\mu(U_s(w)) \leq c_1\mu(K_w)$$

for any $w \in \Lambda_{s,x}$. Since $\cup_{w \in \Lambda_{s,x}} U_s(w) = U_s(x)$,

$$c_*\mu(U_s(x)) \leq c_* \sum_{w \in \Lambda_{s,x}} \mu(U_s(w)) \leq c_1 \sum_{w \in \Lambda_{s,x}} \mu(K_w) = c_1\mu(K_s(x)).$$

This is the condition $(\mathrm{A})_1$ in Section 1.3. Since both μ and $\mathcal{S}_*$ are elliptic, Theorem 1.3.10 implies the condition $(\mathrm{VD})_0$. Hence by Theorem 1.3.5, we have (a). □

LEMMA 3.5.7. (d) *implies* (c).

PROOF. Note that by the previous lemmas, we have (b) and (a). Since d is adapted to the scale $\mathcal{S}_*$, there exist $n \geq 1$ and $c_1 > 0$ such that $U^{(n)}_{c_1 s}(x) \subseteq B_s(x,d)$ for any $x \in X$ and any $s \in (0,1]$. Therefore by (d),

$$p(t,x,x) \leq \frac{c}{\mu(U^{(n)}_{c_1\sqrt{t}}(x))} \leq \frac{c}{\mu(U_{c_1\sqrt{t}}(x))}.$$

Now the volume doubling property of μ with respect to the scale $\mathcal{S}_*$ immediately implies (c). □

LEMMA 3.5.8. *If* (a) *is satisfied, then there exists $c > 0$ such that $r_w \le c r_v$ for any $s \in (0,1]$ and any $w, v \in \Lambda_s$ with $K_w \cap K_v \neq \emptyset$.*

PROOF. By Theorem 1.3.5, μ is gentle with respect to $\mathcal{S}_*$. Hence there exists $c_1 > 0$ such that $\mu(K_w) \le c\mu(K_v)$ and $r_w\mu(K_w) \le c_1 r_v \mu(K_v)$ for any $s \in (0,1]$ and any $w, v \in \Lambda_s$ with $K_w \cap K_v \neq \emptyset$.This shows that $r_w \le (c_1)^2 r_v$. □

PROPOSITION 3.5.9. *Assume that $(\mathcal{E}, \mathcal{F})$ is recurrent. For any closed subset B of K, there exists $g_B : K \times K \to [0, \infty)$ which has the following properties:*
(GF1) *$g_B(x,y) = g_B(y,x) \le g_B(x,x)$ for any $x, y \in K$.For any $x \in K$, define g_B^x by $g_B^x(y) = g_B(x,y)$. Then $g_B^x \in \mathcal{F}_{X\setminus B}$ and $\mathcal{E}(g_B^x, u) = u(x)$ for any $x \in K$ and any $u \in \mathcal{F}_{X\setminus B}$.*
(GF2) *$|g_B(x,y) - g_B(x,z)| \le R(y,z)$ for any $y, z \in K$.*
(GF3) *For $x \notin B$, define*

$$R(x,B) = \big(\min\{\mathcal{E}(u,u) | u \in \mathcal{F}_{X\setminus B}, u(x) = 1\}\big)^{-1}.$$

Then $g_B(x,x) = R(x,B) > 0$.
(GF4) *For $x \notin B$,*

$$E_x(h_B) = \int_K g_B(x,y)\mu(dy).$$

g_B is called the B-Green function.

PROOF. Since $(\mathcal{E}, \mathcal{F})$ is recurrent, $(\mathcal{E}, \mathcal{F})$ is a resistance form on K. If B is a finite set, the above results are shown in [**29**]. Generalization to a closed set is straight forward. See [**27**] for details. □

LEMMA 3.5.10. *Assume* (a). *Set $V_s^{(n)}(x) = \mathrm{int}(U_s^{(n)}(x))$ for any $(s,x) \in (0,1] \times K$ and define $\mathcal{E}_{s,x}(\cdot,\cdot)$ by*

$$\mathcal{E}_{s,x}(u,u) = \sum_{v \in \Lambda_{s,x}^n} \mathcal{E}(u \circ F_v, u \circ F_v)$$

for $u \in \mathcal{F}_{V_s^{(n)}(x)}$.Then there exist $c_1, c_2 > 0$ such that

$$\frac{c_1}{r_w}\mathcal{E}_{s,x}(u,u) \le \mathcal{E}_{V_s^{(n)}(x)}(u,u) \le \frac{c_2}{r_w}\mathcal{E}_{s,x}(u,u)$$

for any $(s,x) \in (0,1] \times K$ and any $u \in \mathcal{F}_{V_s^{(n)}(x)}$.

PROOF. By (SSH), if $u \in \mathcal{F}_{V_s^{(n)}(x)}$,

$$\mathcal{E}_{V_s^{(n)}(x)}(u,u) = \sum_{v \in \Lambda_{s,x}^n} \frac{1}{r_v}\mathcal{E}(u \circ F_v, u \circ F_v)$$

Using Lemma 3.5.8, we immediately deduce the desired inequality. □

LEMMA 3.5.11. *Assume* (a) *and that $(\mathcal{E}, \mathcal{F})$ is recurrent. For $(s,x) \in (0,1] \times K$, define*

$$\overline{R}_{s,x} = \sup_{y \in K_s(x)} R_{s,x}(y) \quad \text{and} \quad \underline{R}_{s,x} = \inf_{y \in K_s(x)} R_{s,x}(y),$$

where

$$R_{s,x}(y) = \big(\inf\{\mathcal{E}_{s,x}(u,u) | u \in \mathcal{F}_{V_s^{(n)}(x)}, u(y) = 1\}\big)^{-1}.$$

Then, $0 < \underline{R}_{s,x} \le \overline{R}_{s,x} < +\infty$ and

$$(c_2)^{-1} r_w \underline{R}_{s,x} \le R(x, V_s^{(n)}(x)^c) \le (c_1)^{-1} r_w \overline{R}_{s,x}$$

for any $w \in \Lambda_{s,x}$, where c_1 and c_2 are the same constants as in Lemma 3.5.10

PROOF. By Lemma 3.5.10, for any $w \in \Lambda_{s,x}$,

$$(c_2)^{-1} r_w R_{s,x}(y) \le R(y, V_s^{(n)}(x)^c) \le (c_1)^{-1} r_w R_{s,x}(y). \tag{3.5.1}$$

Since $R(y, V_s^{(n)}(x)^c) = g_{V_s^{(n)}(x)^c}(y,y)$, it follows that

$$c_1 g_{V_s^{(n)}(x)^c}(y,y) \le r_w R_{s,x}(y) \le c_2 g_{V_s^{(n)}(x)^c}(y,y).$$

Note that $g_{V_s^{(n)}(x)^c}(y,y)$ is continuous with respect to y and is positive for any $y \in K_s(x) \subseteq V_s^{(n)}(x)$. Therefore we see that $0 < \underline{R}_{s,x} \le \overline{R}_{s,x} < +\infty$ because $K_s(x)$ is compact. Now the desired result is straight forward from (3.5.1). □

LEMMA 3.5.12. *Assume that $(\mathcal{E}, \mathcal{F})$ is recurrent. If* (a) *holds, then there exist positive constants c_3 and c_4 such that*

$$c_3 r_w \le R(x, V_s^{(n)}(x)^c) \le c_4 r_w \tag{RES}$$

for any $x \in K$, any $s \in (0,1]$ and any $w \in \Lambda_{s,x}$.

PROOF. Suppose that $(s,x) \underset{n+1}{\sim} (t,y)$. Let ψ be the $n+1$-isomorphism between (s,x) and (t,y) and let ϕ be the associated $\mathcal{L}$-similitude between $U_s^{(n+1)}(x)$ and $U_t^{(n+1)}(y)$. Note that $\psi(\Lambda_{s.x}^k) = \Lambda_{t,y}^k$ for $k = 0, 1, \ldots, n+1$, $\phi(U_s^{(n)}(x)) = U_s^{(n)}(y)$ and $\phi(K_s(x)) = K_s(y)$. Since $\phi(\partial U_s^{(n)}(x)) = \partial U_t^{(n)}(y)$, it follows from (SSH) that $\phi_* : \mathcal{F}_{V_t^{(n)}(y)} \to \mathcal{F}_{V_s^{(n)}(x)}$ defined by $\phi_*(u) = u \circ \phi$ is bijective. Moreover,

$$\begin{aligned}\mathcal{E}_{s,x}(\phi_*(u), \phi_*(u)) &= \sum_{v \in \Lambda_{s,x}^n} \mathcal{E}(\phi_*(u) \circ F_v, \phi_*(u) \circ F_v) \\ &= \sum_{v \in \Lambda_{s,x}^n} \mathcal{E}(u \circ F_{\psi(v)}, u \circ F_{\psi(v)}) = \mathcal{E}_{t,y}(u,u).\end{aligned}$$

Hence $R_{s,x}(z) = R_{t,y}(\phi(z))$. So $\overline{R}_{s,x}$ and $\underline{R}_{s,x}$ depend only on the equivalence classes under $\underset{n+1}{\sim}$. By Theorem 1.3.5, (a) implies that $\mathcal{S}_*$ is locally finite. Hence by Theorems 2.2.7 and 2.2.13, the number of equivalence classes under $\underset{n+1}{\sim}$ is finite. Now Lemma 3.5.11 suffices to deduce the lemma. □

LEMMA 3.5.13. *Assume that $(\mathcal{E}, \mathcal{F})$ is recurrent. If* (a) *holds, then there exists $c_5, c_6 > 0$ such that*

$$c_5 s^2 \le E_x(h_{V_s^{(n)}(x)^c}) \le c_6 s^2 \tag{E}$$

for any $(s,x) \in (0,1] \times K$.

PROOF. First we show the upper estimate. By Proposition 3.5.9,

$$E_x(h_{V_s^{(n)}(x)^c}) = \int_{V_s^{(n)}(x)} g_{V_s^{(n)}(x)^c}(x,y)\mu(dy) \le R(x, V_s^{(n)}(x)^c)\mu(V_s^{(n)}(x)).$$

Since μ is gentle with respect to $\mathcal{S}_*$ and $\mathcal{S}_*$ is locally finite, $\mu(V_s^{(n)}(x)) \le c\mu_w$ for $w \in \Lambda_{s,x}$, where c is independent of s, x and w.This along with Lemma 3.5.12 yields

$$E_x(h_{V_s^{(n)}(x)^c}) \le c c_4 r_w \mu_w \le c_6 s^2.$$

For the lower estimate, note that (SSH) implies

$$R(F_v(y), F_v(z)) \le r_v R(y, z) \tag{3.5.2}$$

for any $y, z \in K$ and any $v \in W_*$. (See [**28**, Lemma 3.3.5] for details.) Hence $\sup_{y,z\in K_v} R(y,z) \le M r_v$ for any $v \in W_*$, where $M = \sup_{p,q\in K} R(p,q)$. Choose m so that $M(\max_{i\in S} r_i)^m \le c_3/2$. Then, for any $w \in \Lambda_{s,x}$, there exists $v \in W_m$ such that $x \in K_{wv}$ and $R(x,y) \le c_3 r_w/2$ for any $y \in K_{wv}$. By (GF2) and Lemma 3.5.12,

$$g_{V_s^{(n)}(x)^c}(x,y) \ge g_{V_s^{(n)}(x)^c}(x,x) - R(x,y) \ge c_3 r_w/2$$

for any $y \in K_{wv}$. Therefore,

$$E_x(h_{V_s^{(n)}(x)^c}) \ge \int_{K_{wv}} g_{V_s^{(n)}(x)^c}(x,y)\mu(dy) \ge c_3 r_w \mu(K_{wv})/2.$$

Since μ is elliptic, $\mu(K_{wv}) \ge b\mu(K_w)$, where b is independent of s, x and w. Therefore we obtain the lower estimate. □

LEMMA 3.5.14. *Assume* (a). *For any $(s,x) \in (0,1] \times K$, define $\mu_{s,x}$ by*

$$\mu_{s,x} = \sum_{v\in\Lambda^n_{s,x}} \mu(F_w^{-1}(A \cap K_v))$$

for any Borel set $A \subseteq K$. If μ is self-similar, then there exist $c_5, c_6 > 0$ such that

$$c_5 \mu_w \mu_{s,x}(A) \le \mu(A) \le c_6 \mu_w \mu_{s,x}(A)$$

for any $(s,x) \in (0,1] \times K$, any $w \in \Lambda_{s,x}$ and any Borel set $A \subseteq U_s(x)$.

LEMMA 3.5.15. *For $(s,x,w) \in (0,1] \times K \times \Lambda_{s,x}$, let $\{E_y^{s,x,w}(\cdot)\}_{y\in V_s^{(n)}(x)}$ be the expectation with respect to the diffusion process associated with the local regular Dirichlet form $(r_w\mathcal{E}_{V_s^{(n)}(x)}, \mathcal{F}_{V_s^{(n)}(x)})$ on $L^2(K, \mu_{s,x})$. Define*

$$\underline{E}^{s,x,w} = \inf_{y\in K_s(x)} E_y^{s,x,w}(h_{V_s^{(n)}(x)^c}) \quad \textit{and} \quad \overline{E}^{s,x,w} = \sup_{y\in K_s(x)} E_y^{s,x,w}(h_{V_s^{(n)}(x)^c}).$$

If (a) *is satisfied, then $0 < \underline{E}^{s,x,w} \le \overline{E}^{s,x,w} < +\infty$ and*

$$c_7 \underline{E}^{s,x,w} s^2 \le E_x(h_{V_s^{(n)}(x)^c}) \le c_8 \overline{E}^{s,x,w} s^2,$$

where c_7 and c_8 are independents of (s,x,w).

PROOF. Use $h(t,y,z)$ to denote the heat kernel associated with the Dirichlet form $(r_w\mathcal{E}_{V_s^{(n)}(x)}, \mathcal{F}_{V_s^{(n)}(x)})$ on $L^2(K, \mu_{s,x})$. Recall that $p_{V_s(x)}(t,y,z)$ is the heat kernel associated with the Dirichlet form $(\mathcal{E}_{V_s^{(n)}(x)}, \mathcal{F}_{V_s^{(n)}(x)})$ on $L^2(K,\mu)$. Therefore, by Lemma 3.5.14,

$$\begin{aligned}\int_{V_s^{(n)}(x)} h(\frac{t}{c_5\mu_w r_w}, y, z)\mu_{s,x}(dz) &\le \int_{V_s^{(n)}(x)} p_{V_s^{(n)}(x)}(t,y,z)\mu(dz) \\ &\le \int_{V_s^{(n)}(x)} h(\frac{t}{c_6\mu_w r_w}, y, z)\mu_{s,x}(dz),\end{aligned}$$

where c_5 and c_6 are the same constants as in Lemma 3.5.14. Integrating this on $[0,\infty)$ with respect to t, we obtain

$$c_5\mu_w r_w E_y^{s,x,w}(h_B) \le E_y(h_B) \le c_6 \mu_w r_w E_y^{s,x,w}(h_B), \tag{3.5.3}$$

where $B = V_s^{(n)}(x)^c$. Note that $h(t,y,z)$ has uniform exponential decay for sufficiently large t, i.e. there exist $c, \lambda > 0$ and $t_* > 0$ such that $h(t,y,z) \le ce^{-\lambda t}$

for any y, z and $t \geq t_*$. Hence $\overline{E}^{s,x,w} < +\infty$. By (UPH), $\int_{y \in K_s(x)} E_y(h_B) > 0$. Hence (3.5.3) implies that $\underline{E}^{s,x,w} > 0$. Again using (3.5.3), we obtain the desired inequality. □

LEMMA 3.5.16. *For any $(s,x), (t,y) \in (0,1] \times K$, we write $(s,x) \underset{*}{\sim} (t,y)$ if and only if $(s,x) \underset{n+1}{\sim} (t,y)$ and there exists $c > 0$ such that $r_{\psi(w)} = cr_w$ for any $w \in \Lambda^n_{s,x}$, where $\psi : \Lambda^{n+1}_{s,x} \to \Lambda^{n+1}_{t,y}$ is the $n+1$-isomorphism between (s,x) and (t,y). Assume* (a). *If (II) is satisfied, i.e. μ is self-similar and the resistance scaling ratio is arithmetic on $\mathcal{R}_1$-relations, then $(0,1] \times K/\underset{*}{\sim}$ is finite.*

PROOF. By Theorem 1.3.5, $\mathcal{S}_*$ is locally finite and $\mathcal{S}_* \underset{\mathrm{GE}}{\sim} \mathcal{S}_\mu$, where $\mathcal{S}_\mu$ is the scale induced by μ. Hence by Theorem 1.4.3-(1), $\mathcal{S}_\mu$ is locally finite as well. Let $(X, Y, \varphi, x, y) \in \mathcal{R}_2$, where $\mathcal{R}$ is the relation set of $\mathcal{L}$. If $(X, Y, \varphi, x, y) \in \mathcal{R}_2$, then Theorem 1.6.1 yields that $r_w \mu_w = r_{\varphi(w)} \mu_{\varphi(w)}$ and $\mu_{\varphi(w)} = \mu_{\varphi(w)}$ for any $w \in X$. Hence $r_w = r_{\varphi(w)}$ for any $w \in X$.

Next we show that $\{r_w/r_v | (w,v) \in \mathcal{IP}(\mathcal{L}, \mathcal{S})\}$ is a finite set. At first, let $(w,v) \in \mathcal{IP}(\mathcal{L}, \mathcal{S}, \mathcal{R}_2)$. Then the above discussion along with Lemma 3.5.8 implies that the choice of the values r_w/r_v is finite.If $(w,v) \in \mathcal{IP}(\mathcal{L}, \mathcal{S}, \mathcal{R}_1$, then we also have finite number of choices of r_w/r_v, because $(r_i)_{i \in S}$ is arithmetic on $\mathcal{R}_1$-relation. Hence $\{r_w/r_v | (w,v) \in \mathcal{IP}(\mathcal{L}, \mathcal{S}, \mathcal{R})\}$ is a finite set. Now let $(w,v) \in \mathcal{IP}(\mathcal{L}, \mathcal{S})$. As in the proof of Theorem 2.2.7, we have $\{w(i)\}_{i=1,\ldots,m+1}$ which satisfies $w(1) = w, w(m+1) = v$ and $(w(i), w(i+1)) \in \mathcal{IP}(\mathcal{L}, \mathcal{S}, \mathcal{R})$ for any i. Note that $m+1 \leq \inf_{p \in K} \#(\pi^{-1}(p)) < \infty$. This fact along with the finiteness of $\{r_w/\mathbf{r}_v | (w,v) \in \mathcal{IP}(\mathcal{L}, \mathcal{S}, \mathcal{R})\}$ implies that $\{r_w/r_v | (w,v) \in \mathcal{IP}(\mathcal{L}, \mathcal{S})\}$ is a finite set.

Now by Theorems 2.2.7 and 2.2.13, the number of equivalence classes under $\underset{n+1}{\sim}$ is finite. Since we only have finite number of choices of r_w/r_v for $(w,v) \in \mathcal{IP}(\mathcal{L}, \mathcal{S}_*)$, one equivalent class of $\underset{n+1}{\sim}$ contains finite number of equivalent classes of $\underset{*}{\sim}$. Therefore, $(0,1]/\underset{*}{\sim}$ is a finite set. □

LEMMA 3.5.17. *Under the assumption* (II), (a) *implies* (E).

PROOF. Let $(s,x) \underset{*}{\sim} (t,y)$, let $\psi : \Lambda^{n+1}_{s,x} \to \Lambda^{n+1}_{t,y}$ be the associated $n+1$-isomorphism and let $\phi : U^{(n+1)}_s(x) \to U^{(n+1)}_t(y)$ be the associated similitude. Choose $w \in \Lambda_{s,x}$. Then ϕ gives a natural correspondence between the Dirichlet forms $(r_w \mathcal{E}_{V^{(n)}_s(x)})$ on $L^2(V^{(n)}_s(x), \mu_{s,x})$ and $(r_{\psi(w)} \mathcal{E}_{V^{(n)}_t(y)})$ on $L^2(V^{(n)}_t(y), \mu_{t,y})$. Therefore, $\underline{E}^{s,x,w} = \underline{E}^{t,y,\psi(w)}$ and $\overline{E}^{s,x,w} = \overline{E}^{t,y,\psi(w)}$. Hence Lemmas 3.5.15 combined with 3.5.16 suffices for (E). □

LEMMA 3.5.18. *Assume* (b). *Then* (DLHK) *and* (UHK) *holds with $\beta > 1$. In particular,* (b) *implies* (d).

PROOF. Note that we have (a) as well due to Theorem 2.3.17. Since d is adapted to $\mathcal{S}_*$, $U^{(n)}_{cr}(x) \subseteq B_r(x,d) \subseteq U^{(n)}_{c'r}(x)$. Hence by Lemmas 3.5.13 and 3.5.17, $c_5 c^2 r^2 \leq E_x(h_{B_r(s,d)^c}) \leq c_6 {c'}^2 r^2$. Let $D(\cdot,\cdot) = d(\cdot,\cdot)^\alpha$.Recall that $\beta = 2/\alpha$. Then, we have the exit time estimate with respect to the distance D:

$$a_1 r^\beta \leq E_x(h_{B_r(x,D)^c}) \leq a_2 r^\beta. \tag{3.5.4}$$

Since μ is gentle with respect to $\mathcal{S}_*$ and $\mathcal{S}_*$ is locally finite, there exists $\gamma > 0$ such that

$$\gamma\mu(U_s^{(n)}(x)) \leq \mu(K_w) \leq \mu(U_s^{(n)}(x)) \tag{3.5.5}$$

for any $(s, x) \in (0, 1] \times K$ and any $w \in \Lambda_{s,x}$. Recall that $\Lambda_s(u) = \{w|w \in \Lambda_s, K_w \cap \text{supp}(u) \neq \emptyset\}$ for $u \in \mathcal{F}$. By (3.5.5),

$$\gamma \min_{x \in \text{supp}(u)} \mu(U_s^{(n)}(x)) \leq \min_{w \in \Lambda_s(u)} \mu(K_w) \leq \min_{x \in \text{supp}(u)} \mu(U_s^{(n)}(x)).$$

Combining this with (3.1.2), we obtain the local Nash inequality:

$$\mathcal{E}(u, u) + \frac{a}{r^\beta \inf_{x \in \text{supp}(u)} \mu(B_r(x, D))} ||u||_1^2 \geq \frac{b}{r^\beta} ||u||_2^2$$

for any $u \in \mathcal{F}$ and any $r \in (0, 1]$, where a and b are independent of u.

Now using [**30**, Theorem 2.9 and Theorem 2.13], we see that $\beta > 1$ and obtain (DLHK) and (UHK). □

By the above lemmas, we see that (a), (b), (c) and (d) are all equivalent.

Now the remaining part of a proof is to show (LHK) under given assumptions.

LEMMA 3.5.19. *Assume* (a). *For any $\epsilon > 0$, there exists $\gamma > 0$ such that $R(x, y)\mu(U_s^{(n)}(x)) \leq \epsilon s^2$ for any $x \in K$, any $y \in U_{\gamma s}^{(n)}(x)$ and any $s \in (0, 1]$.*

PROOF. Write $V(s, x) = \mu(U_s^{(n)}(x))$. By (a), $\mathcal{S}_*$ is locally finite and μ is gentle with respect to $\mathcal{S}_*$. Let $(s, x) \in (0, 1] \times K$ and let $w \in \Lambda_{s,x}$. Then $V(s, x) \leq c\mu(K_w)$ and $s^2 \geq cr_w\mu(K_w)$, where c is independent of (s, x, w). Hence,

$$\frac{R(x, y)V(s, x)}{s^2} \leq c' \frac{R(x, y)}{r_w}. \tag{3.5.6}$$

Now since $\mathcal{S}_*$ is elliptic, for any $m \geq 1$, we can choose $\gamma \in (0, 1)$ so that $|v| \geq m$ if $w' \in \Lambda_s$ and $w'v \in \Lambda_{\gamma s}$. For any $y \in U_{\gamma s}^{(n)}(x)$, there exists $\{w(i)v(i)\}_{i=0}^n \in \Lambda_{\gamma s,x}^n$ such that $w(0) = w, w(k) \in \Lambda_{s,x}^n$ and $K_{w(k-1)v(k-1)} \cap K_{w(k)v(k)} \neq \emptyset$ for any $k = 1, \ldots, n$. Since $r_{w'} \leq ar_w$ for any $w' \in U_s^{(n)}(x)$, where a is independent of s, x and w, (3.5.2) shows that

$$R(x, y) \leq R_* \sum_{k=0}^n r_{w(k)}r_{v(k)} \leq a(n + 1)Mr_w(r_*)^m,$$

where $M = \sup_{p,q \in K} R(p, q)$ and $r_* = \max_{i \in S} r_i$. Choosing a sufficiently large m, we verify the statement of the lemma from (3.5.6). □

LEMMA 3.5.20. *Assume* (a). *In the recurrent case,* (LHK) *holds for any geodesic pair for d^α.*

PROOF. Let $p^{t,x}(y) = p(t, x, y)$ for any t, x, y. Then $p^{t,x}$ belongs to the domain of the self-adjoint operator associated with the Dirichlet from $(\mathcal{E}, \mathcal{F})$ on $L^2(K, \mu)$. By the definition (RF4),

$$\begin{aligned} |p(t, x, y) - p(t, x, x)|^2 &\leq \mathcal{E}(p^{t,x}, p^{t,x})R(x, y) \\ &\leq -\frac{\partial p}{\partial t}(2t, x, x)R(x, y) \leq \frac{p(t, x, x)R(x, y)}{t}. \end{aligned} \tag{3.5.7}$$

(This inequality has been obtained in [**14**, Lemma 6.4] and [**21**, Lemma 5.2].) Combining (3.5.7) with (DLHK) and (c), we obtain

$$p(t,x,y) \geq p(t,x,x)\left(1-\sqrt{\frac{R(x,y)}{tp(t,x,x)}}\right)$$
$$\geq \frac{c}{\mu(B_{\sqrt{t}}(x,d))}\left(1-c'\sqrt{\frac{R(x,y)\mu(U_s^{(n)}(x))}{t}}\right)$$

By Lemma 3.5.19, there exists $\gamma > 0$ such that $R(x,y)\mu(U_{\sqrt{t}}^{(n)}(x))/t \leq (c'/2)^2$ for any $y \in U_{\gamma\sqrt{t}}^{(n)}(x)$. Since d is adapted to $\mathcal{S}_*$,

$$p(t,x,y) \geq \frac{c''}{\mu(B_{\sqrt{t}}(x,d))}$$

for any $y \in B_{\delta\sqrt{t}}(x,d)$. Let $D = s^\alpha$. Rewriting this in terms of D, we have

$$p(t,x,y) \geq \frac{c'''}{\mu(B_{t^{1/\beta}}(x,D))} \tag{3.5.8}$$

for any $y \in B_{\delta' t^{1/\beta}}(x,D)$. This is so called the near diagonal lower estimate. Note that we also have the exit time estimate (3.5.4) and the volume doubling property. By the argument of the proof of [**30**, Theorem 2.13], we obtain (LHK) for geodesic pairs. □

REMARK. In [**30**, Theorem 2.13], it is assumed that the distance is a geodesic distance. However, the discussion of the proof of [**30**, Theorem 2.13] can get through if there exists a geodesic between given two points. The constants are determined by those appeared in the near diagonal estimate and the volume doubling property, and hence they do not depends on the points.

Appendix

A. Existence and continuity of a heat kernel

Let (X, d) be a locally compact metric space and let μ be a Radon measure on (X, d). Let $(\mathcal{E}, \mathcal{F})$ be a regular Dirichlet form on $L^2(X, \mu)$. We use H to denote the non-negative self-adjoint operator from $L^2(X, \mu)$ to itself. Also let $\{T_t\}_{t>0}$ be the strongly continuous semigroup associated with H, i.e.$T_t = e^{-tH}$.

DEFINITION A.1. The semigroup $\{T_t\}_{t>0}$ is said to be ultracontractive if and only if T_t can be extended to a bounded operator from $L^2(X, \mu)$ to $L^\infty(X, \mu)$ for any $t > 0$.

Note that T_t is self-adjoint. Using the duality, T_t can be extended to a bounded operator from $L^1(X, \mu)$ to $L^\infty(X, \mu)$ as well if $\{T_t\}_{t>0}$ is ultracontractive.

One of the conditions implying the ultracontractivity is the Nash inequality.

NOTATION. $||\cdot||_p$ is the L^p-norm of $L^p(X, \mu)$ Also.$||A||_{p\to q}$ is the operator norm of a bounded linear operator $A : L^p(X, \mu) \to L^q(X, \mu)$.

THEOREM A.2. *For $\alpha > 0$, the following conditions (1), (2) and (3) are equivalent.*

(1) *There exist positive constants c_1 and $c_2$such that*

$$(\mathcal{E}(u, u) + c_1||u||_2^2)||u||_1^{2/\alpha} \geq c_2||u||_2^{2+4/\alpha} \tag{A.1}$$

for any $u \in \mathcal{F} \cap L^1(X, d) \cap \mathcal{F}$.

(2) *T_t can be extended to a bounded operator from $L^1(X, \mu)$ to $L^\infty(X, \mu)$ and there exist $c > 0$ such that $||T_t||_{1\to\infty} \leq ct^{-\alpha/2}$ for any $t \in (0, 1]$.*

(3) *$\{T_t\}_{t>0}$ is ultracontractive and there exists $c > 0$ such that $||T_t||_{2\to\infty} \leq ct^{-\alpha/4}$ for any $t \in (0, 1]$.*

(A.1) is called the Nash inequality which was introduced in [**36**]. See [**10, 11, 28**] for the proof of Theorem A.2.

If $\mu(X) < +\infty$, then it is known that the ultracontractivity implies the existence of the heat kernel. The next theorem follows from the results in [**11**, Section 2.1].

THEOREM A.3. *Assume that $\mu(X) < +\infty$ and that $\{T_t\}_{t>0}$ is ultracontractive. Then there exists $p : (0, \infty) \times X \times X \to [0, +\infty)$ such that $p \in L^\infty(X^2, \mu \times \mu)$ and*

$$(T_t u)(x) = \int_X p(t, x, y)u(y)\mu(dy)$$

for any $t > 0$, $x \in X$ and $u \in L^2(X, \mu)$. $p(t, x, y)$ is called the heat kernel associated with the Dirichlet form $(\mathcal{E}, \mathcal{F})$ on $L^2(X, \mu)$. Moreover, H has compact resolvent, i.e. $(H + I)^{-1}$ is a compact operator. Let $(\varphi_k)_{k\geq1}$ be a complete orthonormal system of $L^2(X, \mu)$ consisting of the eigenvalues of H. Assume that $H\varphi_k = \lambda_k\varphi_k$

and $0 \leq \lambda_k \leq \lambda_{k+1}$ *for any* $k \geq 1$ *and* $\lambda_k \to \infty$ *as* $k \to \infty$. *Then* $\varphi_k \in L^\infty(X, \mu)$ *for any* k *and*

$$p(t,x,y) = \sum_{n=1}^{\infty} e^{-\lambda_n t}\varphi_n(x)\varphi_n(y), \tag{A.2}$$

where the infinite sum is uniformly convergent on $[T, +\infty) \times X \times X$ *for any* $T > 0$. *In particular, if* $\varphi_k \in C(X, d)$, *then* $p(t, x, y)$ *is jointly continuous, i.e.* $p : (0, 1] \times X \times X$ *is continuous.*

The next theorem gives a sufficient condition for the heat kernel being positive.

THEOREM A.4. *Assume that* $\mu(X) < +\infty$, *that* $\{T_t\}_{t>0}$ *is ultracontractive and that* (X, d) *is arcwise connected. If the heat kernel is jointly continuous and* $(\mathcal{E}, \mathcal{F})$ *is conservative, then* $p(t, x, y) > 0$ *for any* $(t, x, y) \in (0, +\infty) \times X \times X$.

PROOF. Since $\mathcal{E}(1, 1) = 0$, 1 is an eigenfunction of H. Hence by (A.2), $p(t, x, x) > 0$ for any $x \in X$. Fix $x, y \in X$. Note that if $t > s$, then

$$p(t,x,y) = \int_X p(s,x,z)p(t-s,z,y)\mu(dy). \tag{A.3}$$

Assume that $p(s, x, y) > 0$. Since $p(s, x, y)p(t - s, y, y) > 0$, (A.3) implies that $p(t, x, y) > 0$. Hence there exists $t_* \in [0, +\infty]$ such that $p(t, x, y) = 0$ for any $t \in (0, t_*]$ and $p(t, x, y) > 0$ for any $t \in (t_*, +\infty)$. Next we show that $t_* < +\infty$. Since (X, d) is arcwise connected, there exists $\gamma : [0, 1] \to X$ such that γ is continuous, $\gamma(0) = x$ and $\gamma(1) = y$. For any $s \in [0, 1]$, we have an open neighborhood O_s of $\gamma(s)$ that satisfies, $p(1, z, w) > 0$ for any $z, w \in O_s$. Since $\gamma([0, 1])$ is compact, there exits $\{s_i\}_{i=0}^m$ such that $0 = s_0 < s_1 < \ldots < s_{m-1} < s_m = 1$ and $x_i \in O_{s_{i+1}}$ for any $i = 0, 1, \ldots, m - 1$, where $x_i = \gamma(s_i)$. By (A.3),

$$p(m,x,y) = \int_X \cdots \int_X p(1,x,y_1)p(1,y_1,y_2)\ldots p(1,y_{m-1},y)\mu(dy_1)\ldots\mu(dy_{m-1}).$$

Since $p(1, x_i, x_{i+1}) > 0$ for any $i = 0, 1, \ldots, m - 1$, it follows that $p(m, x, y) > 0$. Therefore, $t_* < m$. Now let $\mathbb{H}_R = \{z | z \in \mathbb{C}, \mathrm{Re}(z) > 0\}$. Then the infinite sum

$$\sum_{i=1}^{\infty} e^{-\lambda_n z}\varphi_n(x)\varphi_n(y)$$

is uniformly convergent on $\mathbb{H}_R$. Hence $p(z, x, y)$ is extended to a holomorphic function on $\mathbb{H}_R$. If $t_* > 0$, then $p(t, x, y) = 0$ for any $t \in (0, t*]$. This implies that $p(z, x, y) = 0$ for any $z \in \mathbb{H}_R$. This obviously contradicts the fact that $t_* < +\infty$. Hence $t_* = 0$. □

DEFINITION A.5. Let X be a set. A pair $(\mathcal{E}, \mathcal{F})$ is called a resistance form on X if it satisfies the following conditions (RF1) through (RF5).
(RF1) $\mathcal{F}$ is a linear subspace of $\ell(X)$ containing constants and $\mathcal{E}$ is a non-negative symmetric quadratic form on $\mathcal{F}$. $\mathcal{E}(u, u) = 0$ if and only if u is constant on X.
(RF2) Let $\sim$ be an equivalent relation on $\mathcal{F}$ defined by $u \sim v$ if and only if $u - v$ is constant on X. Then $(\mathcal{F}/\sim, \mathcal{E})$ is a Hilbert space.
(RF3) For any finite subset $V \subset X$ and for any $v \in \ell(V)$, there exists $u \in \mathcal{F}$ such

that $u|_V = v$.
(RF4) For any $p, q \in X$,

$$\sup\left\{\frac{|u(p)-u(q)|^2}{\mathcal{E}(u,u)}\middle| u \in \mathcal{F}, \mathcal{E}(u,u) > 0\right\}$$

is finite. The above supremum is denoted by $R(p,q)$.
(RF5) If $u \in \mathcal{F}$, then $\bar{u} \in \mathcal{F}$ and $\mathcal{E}(\bar{u},\bar{u}) \leq \mathcal{E}(u,u)$, where

$$\bar{u}(x) = \begin{cases} 1 & \text{if } u(x) \geq 1, \\ u(x) & \text{if } 0 \leq u(x) < 1, \\ 0 & \text{if } u(x) < 0. \end{cases}$$

$R(p,q)$ in the above definition is called the effective resistance between p and q. It is known that $R(\cdot,\cdot)$ is a distance on X. We call $R(\cdot,\cdot)$ the resistance metric associated with the resistance form $(\mathcal{E},\mathcal{F})$. See [**28**] and [**29**] for more details on resistance forms.

THEOREM A.6. *Assume that $\mu(X) < +\infty$, that $\{T_t\}_{t>0}$ is ultracontractive and that there exist $\alpha \in (0,2)$ and $c > 0$ such that*

$$||T_t||_{1\to\infty} \leq ct^{-\alpha/2} \tag{A.4}$$

for any $t \in (0,1]$. Then, we may choose $M > 0$ so that

$$\mathcal{E}_*(u,u) \geq M||u||_\infty^2 \tag{A.5}$$

for any $u \in \mathcal{F}$, where $\mathcal{E}_*(u,u) = \mathcal{E}(u,u) + ||u||_2^2$. In particular, $\mathcal{F} \subseteq C(X,d)$ and the heat kernel $p(t,x,y)$ associated with the Dirichlet form $(\mathcal{E},\mathcal{F})$ on $L^2(X,\mu)$ is jointly continuous. Moreover if $(\mathcal{E},\mathcal{F})$ is conservative and there exists $c' > 0$ such that

$$\mathcal{E}(u,u) \geq c' \int_X (u-\bar{u})^2 d\mu \tag{A.6}$$

for any $u \in \mathcal{F}$, where $\bar{u} = \mu(X)^{-1}\int_X u d\mu$, then $(\mathcal{E},\mathcal{F})$ is a resistance form on X. Also if R is the resistance metric associated with $(\mathcal{E},\mathcal{F})$, then (X,R) is bounded.

PROOF. Define $G_* u = \int_0^\infty e^{-t}T_t u d\mu$. By (A.4), $\int_0^\infty e^{-t}||T_t||_{1\to\infty}dt < +\infty$. Hence $G_* : L^1(X,\mu) \to L^\infty(X,\mu)$ is a bounded operator. Since $G_*\varphi_k = (\lambda_k + 1)^{-1}\varphi_k$ for any $k \geq 1$, we have $G_*|_{L^2(X,\mu)} = (H+I)^{-1}$. Note that $\mathcal{E}_*(G_*u, G_*u) = (u, G_*u)$ for any $u \in L^2(X,\mu)$. Hence,

$$||G_*u||_*^2 \leq ||u||_1||G_*u||_\infty \leq M||u||_1^2,$$

where $||v||_* = \sqrt{\mathcal{E}_*(v,v)}$ and $M = ||G_*||_{1\to\infty}$. Now $\mathcal{E}_*(u, G_*v) = (u,v)$ for any $u \in \mathcal{F}$ and any $v \in L^2(X,\mu)$. Therefore

$$|(v,u)| \leq |\mathcal{E}_*(u,G_*v)| \leq M||u||_*||G_*u||_* \leq \sqrt{M}||u||_*||v||_1.$$

Since $L^2(X,\mu)$ is dense in $L^1(X,\mu)$, we have $u \in L^\infty(X,\mu)$ and (A.5).Since $(\mathcal{E},\mathcal{F})$ is regular, there exist a core $C \subseteq \mathcal{F} \cap C_0(X,d)$ such that C is dense in $C_0(X,d)$ with respect to $||\cdot||_\infty$ and in $\mathcal{F}$ with respect to $||\cdot||_*$. By (A.5), it follows that $\mathcal{F} \subseteq C(X,d)$. Now that $\varphi_k \in C(X,d)$, Theorem A.3 shows the continuity of the heat kernel.

Next we will verify the conditions (RF1) - (RF5) to show that $(\mathcal{E}, \mathcal{F})$ is a resistance form of X. (RF1) is immediate from the fact that $1 \in \mathcal{F}$, $\mathcal{E}(1,1) = 0$ and (A.6). By (A.6) and (A.5),

$$(A.7) \quad (1+c')\mathcal{E}(u,u) \geq \mathcal{E}(u-\bar{u}, u-\bar{u}) + ||u-\bar{u}||_2^2 = ||u-\bar{u}||_*^2 \geq M||u-\bar{u}||_\infty^2$$

If $\mathcal{F}_* = \{u | u \in \mathcal{F}, \bar{u} = 0\}$, then (A.7) says that $\mathcal{E}$ and $\mathcal{E}_*$ are equivalent on $\mathcal{F}_*$. Since $(\mathcal{F}, \mathcal{E}_*)$ is complete, $(\mathcal{F}_*, \mathcal{E})$ is complete. This implies (RF2). Again by (A.7), there exists $c_1 > 0$ such that

$$c_1\mathcal{E}(u,u) \geq c_1||u-\bar{u}||_\infty^2$$

for any $u \in \mathcal{F}$. Therefore, for any $p, q \in X$ and any $u \in \mathcal{F}$,

$$|u(p) - u(q)|^2 \leq (|u(p) - \bar{u}| + |u(q) - \bar{u}|)^2 \leq 2c_1\mathcal{E}(u,u).$$

Hence

$$(A.8) \qquad \sup\left\{ \frac{|u(p) - u(q)|^2}{\mathcal{E}(u,u)} \middle| \mathcal{E}(u,u) > 0 \right\} \leq 2c_1$$

for any $p, q \in X$. So we have (RF4). (RF3) holds because $\mathcal{F}$ is dense in $C_0(X,d)$. (RF5) is immediate from the Markov property of $(\mathcal{E}, \mathcal{F})$. Thus we obtain the conditions (RF1) through (RF5). Finally by (A.8), $\sup_{p,q \in X} R(p,q) \leq 2c_1$. □

B. Recurrent case and resistance form

Let $(K, S, \{F_i\}_{i \in S})$ be a self-similar structure and let d be a metric on K which gives the natural topology of K associated with the self-similar structure. We will consider a resistance from $(\mathcal{E}, \mathcal{F})$ on K which satisfies the following conditions (RFA1), (RFA2) and (RFA3):

(RFA1) $u \circ F_i \in \mathcal{F}$ for any $i \in S$. Moreover there exists $(r_i)_{i \in S} \in (0,1)^S$ such that

$$\mathcal{E}(u,v) = \sum_{i \in S} \frac{1}{r_i} \mathcal{E}(u \circ F_i, v \circ F_i)$$

for any $u, v \in \mathcal{F}$.

(RFA2) Let R be the resistance metric on K associated with $(\mathcal{E}, \mathcal{F})$. Then (K, R) is bounded.

(RFA3) $\mathcal{F} \subseteq C(K,d)$ and $\mathcal{F}$ is dense in $C(K,d)$.

PROPOSITION B.1. *Under the above situation, R gives the same topology as the one given by d.*

PROOF. Using the same arguments as in [**28**, Lemma 3.3.5], we have

$$(B.1) \qquad r_w R(p,q) \geq R(F_w(p), F_w(q))$$

for any $w \in W_*$ and any $p, q \in K$.

Let $R(x_n, x) \to 0$ as $n \to \infty$. Since (K,d) is compact, there exists $x_* \in K$ such that $d(x_{n_i}, x_*) \to 0$ as $i \to \infty$ for some $\{n_i\}_i$. Since $f \in C(K,d) \cap C(K,R)$ for any $f \in \mathcal{F}$, we see that $f(x) = \lim_{i \in \infty} f(x_{n_i}) = f(x_*)$. Hence $x = x_*$ because $(\mathcal{E}, \mathcal{F})$ is a resistance form. This implies that $d(x_n, x) \to 0$ as $n \to \infty$.

Conversely, assume $\lim_{n \to \infty} d(x_n, x) = 0$. Define $K_m(x) = \cup_{w \in W_m : x \in K_w} K_w$. Then for any $m \geq 0$, $x_n \in K_m(x)$ for sufficiently large n. Hence (B.1) along with (RFA2) implies that $R(x_n, x) \leq r_w(\sup_{p,q \in X} R(p,q))$ if $w \in W_m$ and $x \in K_w$. Therefore $R(x_n, x) \to 0$ as $n \to \infty$. □

LEMMA B.2. *Assume* (RFA1), (RFA2) and (RFA3). *There exists* $c > 0$ *such that*

(B.2)
$$\mathcal{E}(u,u) \geq c \int_K (u - (\bar{u})_\mu)^2 d\mu$$

for any $u \in \mathcal{F}$ *and any elliptic probability measure* μ *on* K, *where* $(\bar{u})_\mu = \int_X u d\mu$.

PROOF. (RFA2) implies that $M = \sup_{p,q \in K} R(p,q)$ is finite. Then,

$$M\mathcal{E}(u,u) \geq R(p,q)\mathcal{E}(u,u) \geq |u(p) - u(q)|^2$$

for any $u \in \mathcal{F}$ and any $p, q \in K$. Integrating this with respect to p and q, we immediately obtain (B.2). □

THEOREM B.3. *Let* μ *be an admissible measure on* (K,d). *The following two conditions* (RE1) *and* (RE2) *are equivalent.*

(RE1) μ *is elliptic.* $(\mathcal{E}, \mathcal{F})$ *is a resistance form on* K *which satisfies* (RFA1), (RFA2) *and* (RFA3).

(RE2) $(\mathcal{E}, \mathcal{F})$ *is a local regular Dirichlet form on* $L^2(K,\mu)$. $1 \in \mathcal{F}$ *and* $\mathcal{E}(1,1) = 0$. $(\mathcal{E}, \mathcal{F}, \mu)$ *satisfies* (SSF) *and* (PI) *and is recurrent.*

Moreover if (RE1) *or* (RE2) *holds, then* (CHK) *and* (UPH) *are satisfied.*

PROOF. Note that both (RE1) and (RE2) implies $0 < r_i < 1$ for any $i \in S_*$. Therefore, if μ is elliptic, than $\mathcal{S}_*$ is elliptic as well.

First we assume (RE1). By [**28**, Theorem 2.4.2], $(\mathcal{E}, \mathcal{F})$ is a regular Dirichlet from on $L^2(K,\mu)$. To show the local property, suppose that $u, v \in \mathcal{F}$ and $\mathrm{supp}(u) \cap \mathrm{supp}(v) = \emptyset$. Then we may choose m so that $K_w \cap \mathrm{supp}(u) \cap \mathrm{supp}(v) = \emptyset$ for any $w \in W_m$. Then by (RFA1), $\mathcal{E}(u,v) = \sum_{w \in W_m} (r_w)^{-1} \mathcal{E}(u \circ F_w, v \circ F_w) = 0$. Hence $(\mathcal{E}, \mathcal{F})$ has the local property. (SSF) is immediate form (RFA1). (PI) follows from Lemma B.2.

Conversely, assume (RE2). Then by Theorem 3.1.4, we have all the properties required in Theorem A.6. Therefore, we have (RE1).

Finally if (RE2) holds, then by Theorem 3.1.8, we have (CHK) and (UPH). □

C. Heat kernel estimate to the volume doubling property

In this section, (X, d) is a locally compact metric space where every bounded set is precompact, μ is a Radon measure on (X,d) and $(\mathcal{E}, \mathcal{F})$ is a local regular Dirichlet form on $L^2(X,\mu)$. Let H be the non-negative self-adjoint operator on $L^2(X,\mu)$ associated with $(\mathcal{E}, \mathcal{F})$ and let $\{T_t\}_{t>0}$ be the strongly continuous semigroup on $L^2(X,\mu)$ associated with H.Also let $(\{X_t\}_{t>0}, \{P_x\}_{x \in X})$ be the diffusion process associated with $(\mathcal{E}, \mathcal{F})$. We assume that $\{T_t\}_{t>0}$ is ultracontractive.

Let U be a nonempty open subset of X and let μ_U be the restriction of μ on U. Define $\mathcal{D}_U = \{u | u \in \mathcal{F} \cap C(X), u|_{X \setminus U} \equiv 0\}$. Let $\mathcal{F}_U$ be the closure of $\mathcal{D}_U$ with respect to the inner product $\mathcal{E}_*(u,v) = \mathcal{E}(u,v) + \int_X uv d\mu$ and let $\mathcal{E}_U = \mathcal{E}|_{\mathcal{F}_U \times \mathcal{F}_U}$. By [**15**, Theorem 4.3], $(\mathcal{E}_U, \mathcal{F}_U)$ is a local regular Dirichlet form on $L^2(U, \mu_U)$. Moreover, if $(\{X_t^U\}_{t>0}, \{P_x^U\}_{x \in U})$ is the diffusion process associated with $(\mathcal{E}_U, \mathcal{F}_U)$ and $\tau_U = \inf\{t | X_t \notin U\}$, then

(C.1)
$$P_x^U(X_t^U \in A) = P_x(X_t \in A, \tau_U \geq t\}.$$

PROPOSITION C.1. *Let* $\{T_t^U\}_{t>0}$ *be the strongly continuous semigroup associated with* $(\mathcal{E}, \mathcal{F})$. *Then* $\{T_t^U\}_{t>0}$ *is ultracontractive.*

PROOF. By (C.1),if $u \geq 0$, then

$$(T_t^U u)(x) \leq (T_t u)(x) \tag{C.2}$$

for μ-a.e. $x \in X$. This immediately shows the desired statement. □

DEFINITION C.2. Let U be a nonempty open subset of X.Define $\lambda_*(U)$ be

$$\lambda_*(U) = \inf_{u \in \mathcal{F}_U, u \neq 0} \frac{\mathcal{E}_U(u,u)}{||u||_2^2}.$$

By the variational formula, $\lambda_*(U)$ is the bottom of the spectrum of $-\Delta_U$, where $-\Delta_U$ is the non-negative self-adjoint operator on $L^2(U, \mu_U)$ associated with $(\mathcal{E}_U, \mathcal{F}_U)$.

THEOREM C.3. *Assume that the heat kernel $p(t,x,y)$ associated with $(\mathcal{E}, \mathcal{F})$ is jointly continuous. Suppose that the following two conditions* (RFK) *and* (DUHK) *are satisfied for some $\beta > 0$:*
(RFK) *There exist $r_* > 0$ and $c_1 > 0$ such that*

$$\lambda_*(B_r(x)) \leq c_1 r^{-\beta}$$

for any $r \in (0, r_*]$ and any $x \in X$.
(DUHK) *There exist positive constants t_*, c_2 and c_3 such that*

$$p(t,x,x) \leq \frac{c_2}{\mu(B_{c_3 t^{1/\beta}}(x))}$$

for any $t \in (0, t_*]$ and any $x \in X$.
Then for any $r \in (0, \min\{c_3(t_*)^{1/\beta}/3, r_*\}]$,

$$\mu(B_{2r}(x)) \leq c\mu(B_r(x)),$$

where $c > 0$ is a constant which is independent of x and r.

PROOF. Let $r \in (0, \min\{c_3(t_*)^{1/\beta}/3, r_*\}]$. For any $y \in B_r(x)$, (DUHK) implies that

$$p(c_* r^\beta, y, y) \leq \frac{c_2}{\mu(B_{3r}(y))} \leq \frac{c_2}{\mu(B_{2r}(x))}, \tag{C.3}$$

where $c_* = (3/c_3)^\beta$. Note that $\mu(B_r(x)) < +\infty$. Hence by Theorem A.3, there exists a heat kernel $p_{B_r(x)}(t,y,z)$ associated with $(\mathcal{E}_{B_r(x)}, \mathcal{F}_{B_r(x)})$.Using (C.2), we see that

$$p_{B_r(x)}(t,y,z) \leq p(t,y,z)$$

for $\mu \times \mu$-a.e. $(y,z) \in X^2$. Therefore,

$$\begin{aligned} e^{-\lambda_*(B_r(x))t} &\leq \sum_{i \geq 1} e^{-\lambda_i t} = \int_{B_r(x)^2} p_{B_r(x)}(t/2, y, z)^2 \mu(dy)\mu(dz) \\ &\leq \int_{B_r(x) \times X} p(t/2, y, z)^2 \mu(dy)\mu(dz) = \int_{B_r(x)} p(t,y,y)\mu(dy), \end{aligned}$$

where $\{\lambda_i\}_{i \geq 1}$ be the eigenvalues of $-\Delta_U$. This and (C.3) along with (RFK) show that

$$e^{-c_1 c_*} \leq e^{-\lambda_*(B_r(x))c_* r^\beta} \leq c_2 \frac{\mu(B_r(x))}{\mu(B_{2r}(x))}.$$

□

Bibliography

[1] M. T. Barlow, *Diffusion on fractals*, Lecture notes Math. vol. 1690, Springer, 1998.

[2] M. T. Barlow and R. F. Bass, *The construction of Brownian motion on the Sierpinski carpet*, Ann. Inst. Henri Poincaré **25** (1989), 225–257.

[3] ———, *Local time for Brownian motion on the Sierpinski carpet*, Probab. Theory Related Fields **85** (1990), 91–104.

[4] ———, *On the resistance of the Sierpinski carpet*, Proc. R. Soc. London A **431** (1990), 354–360.

[5] ———, *Transition densities for Brownian motion on the Sierpinski carpet*, Probab. Theory Related Fields **91** (1992), 307–330.

[6] ———, *Coupling and Harnack inequalities for Sierpinski carpets*, Bull. Amer. Math. Soc. (N. S.) **29** (1993), 208–212.

[7] ———, *Brownian motion and harmonic analysis on Sierpinski carpets*, Canad. J. Math. **51** (1999), 673–744.

[8] M. T. Barlow and T. Kumagai, *Transition density asymptotics for some diffusion processes with multi-fractal structures*, Electron. J. Probab. **6** (2001), 1–23.

[9] M. T. Barlow and E. A. Perkins, *Brownian motion on the Sierpinski gasket*, Probab. Theory Related Fields **79** (1988), 542–624.

[10] E. Carlen, S. Kusuoka, and D. Stroock, *Upper bounds for symmetric Markov transition functions*, Ann. Inst. Henri Poincaré **23** (1987), 245–287.

[11] E. B. Davies, *Heat Kernels and Spectral Theory*, Cambridge Tracts in Math. vol 92, Cambridge University Press, 1989.

[12] K. J. Falconer, *Fractal Geometry*, Wiley, 1990.

[13] ———, *Techniques in Fractal Geometry*, Wiley, 1997.

[14] P. J. Fitzsimmons, B. M. Hambly, and T. Kumagai, *Transition density estimates for Brownian motion on affine nested fractals*, Comm. Math. Phys. **165** (1994), 595–620.

[15] M. Fukushima, Y. Oshima, and M. Takeda, *Dirichlet Forms and Symmetric Markov Processes*, de Gruyter Studies in Math. vol. 19, de Gruyter, Berlin, 1994.

[16] A. Grigor'yan, *The heat equation on noncompact Riemannian manifolds. (in Russian)*, Mat. Sb. **182** (1991), 55–87, English translation in Math. USSR-Sb. 72(1992), 47–77.

[17] A. Grigor'yan and A. Telcs, in preparation.

[18] ———, *Sub-Gaussian estimates of heat kernels on infinite graphs*, Duke Math. J. **109** (2001), 451 - 510.

[19] ———, *Harnack inequalities and sub-Gaussian estimates for random walks*, Math. Ann. **324** (2002), 521–556.

[20] B. M. Hambly, J. Kigami, and T. Kumagai, *Multifractal formalisms for the local spectral and walk dimensions*, Math. Proc. Cambridge Phil. Soc. **132** (2002), 555–571.

[21] B. M. Hambly and T. Kumagai, *Transition density estimates for diffusion processes on post critically finite self-similar fractals*, Proc. London Math. Soc. (3) **78** (1999), 431–458.

[22] B. M. Hambly, T. Kumagai, S. Kusuoka, and X. Y. Zhou, *Transition density estimates for diffusion processes on homogeneous random Sierpinski carpets*, J. Math. Soc. Japan **52** (2000), 373–408.

[23] J. Heinonen, *Lectures on Analysis on Metric Spaces*, Springer, 2001.

[24] J. E. Hutchinson, *Fractals and self similarity*, Indiana Univ. Math. J. **30** (1981), 713–747.

[25] A. Kameyama, *Self-similar sets from the topological point of view*, Japan J. Indust. Appl. Math. **10** (1993), 85–95.

[26] ———, *Distances on topological self-similar sets and the kneading determinants*, J. Math. Kyoto Univ. **40** (2000), 601–672.

[27] J. Kigami, in preparation.
[28] ———, *Analysis on Fractals*, Cambridge Tracts in Math. vol. 143, Cambridge University Press, 2001.
[29] ———, *Harmonic analysis for resistance forms*, J. Functional Analysis **204** (2003), 399–444.
[30] ———, *Local Nash inequality and inhomogeneity of heat kernels*, Proc. London Math. Soc. (3) **89** (2004), 525–544.
[31] J. Kigami, R. S. Strichartz, and K. C. Walker, *Constructing a Laplacian on the diamond fractal*, Experimental Math. **10** (2001), 437–448.
[32] H. Kimura, *Self-similar geodesic distances on the sierpinski carpet*, Master thesis, Kyoto University, 2005, in Japanese.
[33] T. Kumagai, *Estimates of the transition densities for Brownian motion on nested fractals*, Probab. Theory Related Fields **96** (1993), 205–224.
[34] S. Kusuoka and X. Y. Zhou, *Dirichlet forms on fractals: Poincaré constant and resistance*, Probab. Theory Related Fields **93** (1992), 169–196.
[35] P. A. P. Moran, *Additive functions of intervals and Hausdorff measure*, Proc. Cambridge Phil. Soc. **42** (1946), 15–23.
[36] J. Nash, *Continuity of solutions of parabolic and elliptic equations*, Amer. J. Math. **80** (1958), 931–954.
[37] R. T. Rockafeller, *Convex Analysis*, Princeton Univ. Press, 1970.
[38] L. Saloff-Coste, *A note on Poincaré, Sobolev, and Harnack inequalities*, Internat. Math. Res. Notices (1992), 27–38.
[39] E. Stiemke, *Über positive Lösungen homogener linearer Geleichhungen*, Math. Ann. **76** (1915), 340–342.

Assumptions, Conditions and Properties in Parentheses

List of Notations

Index

Editorial Information

To be published in the *Memoirs*, a paper must be correct, new, nontrivial, and significant. Further, it must be well written and of interest to a substantial number of mathematicians. Piecemeal results, such as an inconclusive step toward an unproved major theorem or a minor variation on a known result, are in general not acceptable for publication.

Papers appearing in *Memoirs* are generally at least 80 and not more than 200 published pages in length. Papers less than 80 or more than 200 published pages require the approval of the Managing Editor of the Transactions/Memoirs Editorial Board.

As of January 31, 2009, the backlog for this journal was approximately 11 volumes. This estimate is the result of dividing the number of manuscripts for this journal in the Providence office that have not yet gone to the printer on the above date by the average number of monographs per volume over the previous twelve months, reduced by the number of volumes published in four months (the time necessary for preparing a volume for the printer). (There are 6 volumes per year, each usually containing at least 4 numbers.)

A Consent to Publish and Copyright Agreement is required before a paper will be published in the *Memoirs*. After a paper is accepted for publication, the Providence office will send a Consent to Publish and Copyright Agreement to all authors of the paper. By submitting a paper to the *Memoirs*, authors certify that the results have not been submitted to nor are they under consideration for publication by another journal, conference proceedings, or similar publication.

Information for Authors

Memoirs are printed from camera copy fully prepared by the author. This means that the finished book will look exactly like the copy submitted.

Initial submission. The AMS uses Centralized Manuscript Processing for initial submissions. Authors should submit a PDF file using the Initial Manuscript Submission form found at `www.ams.org/peer-review-submission`, or send one copy of the manuscript to the following address: Centralized Manuscript Processing, MEMOIRS OF THE AMS, 201 Charles Street, Providence, RI 02904-2294 USA. If a paper copy is being forwarded to the AMS, indicate that it is for it Memoirs and include the name of the corresponding author, contact information such as email address or mailing address, and the name of an appropriate Editor to review the paper (see the list of Editors below).

The paper must contain a *descriptive title* and an *abstract* that summarizes the article in language suitable for workers in the general field (algebra, analysis, etc.). The *descriptive title* should be short, but informative; useless or vague phrases such as "some remarks about" or "concerning" should be avoided. The *abstract* should be at least one complete sentence, and at most 300 words. Included with the footnotes to the paper should be the 2000 *Mathematics Subject Classification* representing the primary and secondary subjects of the article. The classifications are accessible from `www.ams.org/msc/`. The list of classifications is also available in print starting with the 1999 annual index of *Mathematical Reviews*. The Mathematics Subject Classification footnote may be followed by a list of *key words and phrases* describing the subject matter of the article and taken from it. Journal abbreviations used in bibliographies are listed in the latest *Mathematical Reviews* annual index. The series abbreviations are also accessible from `www.ams.org/msnhtml/serials.pdf`. To help in preparing and verifying references, the AMS offers MR Lookup, a Reference Tool for Linking, at `www.ams.org/mrlookup/`.

Electronically prepared manuscripts. The AMS encourages electronically prepared manuscripts, with a strong preference for $\mathcal{A}_{\mathcal{M}}\mathcal{S}$-LaTeX. To this end, the Society has prepared $\mathcal{A}_{\mathcal{M}}\mathcal{S}$-LaTeX author packages for each AMS publication. Author packages include instructions for preparing electronic manuscripts, samples, and a style file that generates the particular design specifications of that publication series. Though $\mathcal{A}_{\mathcal{M}}\mathcal{S}$-LaTeX is the highly preferred format of TeX, author packages are also available in $\mathcal{A}_{\mathcal{M}}\mathcal{S}$-TeX.

Authors may retrieve an author package for *Memoirs of the AMS* from `www.ams.org/journals/memo/memoauthorpac.html` or via FTP to `ftp.ams.org` (login as `anonymous`, enter username as password, and type `cd pub/author-info`). The *AMS Author Handbook* and the *Instruction Manual* are available in PDF format from the author package link. The author package can also be obtained free of charge by sending email to `tech-support@ams.org` (Internet) or from the Publication Division, American Mathematical Society, 201 Charles St., Providence, RI 02904-2294, USA. When requesting an author package, please specify $\mathcal{A}_{\mathcal{M}}\mathcal{S}$-LaTeX or $\mathcal{A}_{\mathcal{M}}\mathcal{S}$-TeX and the publication in which your paper will appear. Please be sure to include your complete mailing address.

After acceptance. The final version of the electronic file should be sent to the Providence office (this includes any TeX source file, any graphics files, and the DVI or PostScript file) immediately after the paper has been accepted for publication.

Before sending the source file, be sure you have proofread your paper carefully. The files you send must be the EXACT files used to generate the proof copy that was accepted for publication. For all publications, authors are required to send a printed copy of their paper, which exactly matches the copy approved for publication, along with any graphics that will appear in the paper.

Accepted electronically prepared files can be submitted via the web at `www.ams.org/submit-book-journal/`, sent via FTP, or sent on CD-Rom or diskette to the Electronic Prepress Department, American Mathematical Society, 201 Charles Street, Providence, RI 02904-2294 USA. TeX source files, DVI files, and PostScript files can be transferred over the Internet by FTP to the Internet node `ftp.ams.org` (130.44.1.100). When sending a manuscript electronically via CD-Rom or diskette, please be sure to include a message identifying the paper as a Memoir.

Electronically prepared manuscripts can also be sent via email to `pub-submit@ams.org` (Internet). In order to send files via email, they must be encoded properly. (DVI files are binary and PostScript files tend to be very large.)

Electronic graphics. Comprehensive instructions on preparing graphics are available at `www.ams.org/authors/journals.html`. A few of the major requirements are given here.

Submit files for graphics as EPS (Encapsulated PostScript) files. This includes graphics originated via a graphics application as well as scanned photographs or other computer-generated images. If this is not possible, TIFF files are acceptable as long as they can be opened in Adobe Photoshop or Illustrator. No matter what method was used to produce the graphic, it is necessary to provide a paper copy to the AMS.

Authors using graphics packages for the creation of electronic art should also avoid the use of any lines thinner than 0.5 points in width. Many graphics packages allow the user to specify a "hairline" for a very thin line. Hairlines often look acceptable when proofed on a typical laser printer. However, when produced on a high-resolution laser imagesetter, hairlines become nearly invisible and will be lost entirely in the final printing process.

Screens should be set to values between 15% and 85%. Screens which fall outside of this range are too light or too dark to print correctly. Variations of screens within a graphic should be no less than 10%.

Inquiries. Any inquiries concerning a paper that has been accepted for publication should be sent to `memo-query@ams.org` or directly to the Electronic Prepress Department, American Mathematical Society, 201 Charles St., Providence, RI 02904-2294 USA.

Editors

This journal is designed particularly for long research papers, normally at least 80 pages in length, and groups of cognate papers in pure and applied mathematics. Papers intended for publication in the *Memoirs* should be addressed to one of the following editors. The AMS uses Centralized Manuscript Processing for initial submissions to AMS journals. Authors should follow instructions listed on the Initial Submission page found at `www.ams.org/memo/memosubmit.html`.

Algebra to ALEXANDER KLESHCHEV, Department of Mathematics, University of Oregon, Eugene, OR 97403-1222; email: `ams@noether.uoregon.edu`

Algebraic geometry to DAN ABRAMOVICH, Department of Mathematics, Brown University, Box 1917, Providence, RI 02912; email: `amsedit@math.brown.edu`

Algebraic geometry and its applications to MINA TEICHER, Emmy Noether Research Institute for Mathematics, Bar-Ilan University, Ramat-Gan 52900, Israel; email: `teicher@macs.biu.ac.il`

Algebraic topology to ALEJANDRO ADEM, Department of Mathematics, University of British Columbia, Room 121, 1984 Mathematics Road, Vancouver, British Columbia, Canada V6T 1Z2; email: `adem@math.ubc.ca`

Combinatorics to JOHN R. STEMBRIDGE, Department of Mathematics, University of Michigan, Ann Arbor, Michigan 48109-1109; email: `JRS@umich.edu`

Commutative and homological algebra to LUCHEZAR L. AVRAMOV, Department of Mathematics, University of Nebraska, Lincoln, NE 68588-0130; email: `avramov@math.unl.edu`

Complex analysis and harmonic analysis to ALEXANDER NAGEL, Department of Mathematics, University of Wisconsin, 480 Lincoln Drive, Madison, WI 53706-1313; email: `nagel@math.wisc.edu`

Differential geometry and global analysis to CHRIS WOODWARD, Department of Mathematics, Rutgers University, 110 Frelinghuysen Road, Piscataway, NJ 08854; email: `ctw@math.rutgers.edu`

Dynamical systems and ergodic theory and complex analysis to YUNPING JIANG, Department of Mathematics, CUNY Queens College and Graduate Center, 65-30 Kissena Blvd., Flushing, NY 11367; email:CcCC `Yunping.Jiang@qc.cuny.edu`

Functional analysis and operator algebras to DIMITRI SHLYAKHTENKO, Department of Mathematics, University of California, Los Angeles, CA 90095; email: `shlyakht@math.ucla.edu`

Geometric analysis to WILLIAM P. MINICOZZI II, Department of Mathematics, Johns Hopkins University, 3400 N. Charles St., Baltimore, MD 21218; email: `trans@math.jhu.edu`

Geometric topology to MARK FEIGHN, Math Department, Rutgers University, Newark, NJ 07102; email: `feighn@andromeda.rutgers.edu`

Harmonic analysis, representation theory, and Lie theory to ROBERT J. STANTON, Department of Mathematics, The Ohio State University, 231 West 18th Avenue, Columbus, OH 43210-1174; email: `stanton@math.ohio-state.edu`

Logic to STEFFEN LEMPP, Department of Mathematics, University of Wisconsin, 480 Lincoln Drive, Madison, Wisconsin 53706-1388; email: `lempp@math.wisc.edu`

Number theory to JONATHAN ROGAWSKI, Department of Mathematics, University of California, Los Angeles, CA 90095; email: `jonr@math.ucla.edu`

Number theory to SHANKAR SEN, Department of Mathematics, 505 Malott Hall, Cornell University, Ithaca, NY 14853; email: `ss70@cornell.edu`

Partial differential equations to GUSTAVO PONCE, Department of Mathematics, South Hall, Room 6607, University of California, Santa Barbara, CA 93106; email: `ponce@math.ucsb.edu`

Partial differential equations and dynamical systems to PETER POLACIK, School of Mathematics, University of Minnesota, Minneapolis, MN 55455; email: `polacik@math.umn.edu`

Probability and statistics to RICHARD BASS, Department of Mathematics, University of Connecticut, Storrs, CT 06269-3009; email: `bass@math.uconn.edu`

Real analysis and partial differential equations to DANIEL TATARU, Department of Mathematics, University of California, Berkeley, Berkeley, CA 94720; email: `tataru@math.berkeley.edu`

All other communications to the editors should be addressed to the Managing Editor, ROBERT GURALNICK, Department of Mathematics, University of Southern California, Los Angeles, CA 90089-1113; email: `guralnic@math.usc.edu`.

Titles in This Series

935 **Mihai Ciucu,** The scaling limit of the correlation of holes on the triangular lattice with periodic boundary conditions, 2009

934 **Arjen Doelman, Björn Sandstede, Arnd Scheel, and Guido Schneider,** The dynamics of modulated wave trains, 2009

933 **Luchezar Stoyanov,** Scattering resonances for several small convex bodies and the Lax-Phillips conjecture, 2009

932 **Jun Kigami,** Volume doubling measures and heat kernel estimates on self-similar sets, 2009

931 **Robert C. Dalang and Marta Sanz-Solé,** Hölder-Sobolev regularity of the solution to the stochastic wave equation in dimension three, 2009

930 **Volkmar Liebscher,** Random sets and invariants for (type II) continuous tensor product systems of Hilbert spaces, 2009

929 **Richard F. Bass, Xia Chen, and Jay Rosen,** Moderate deviations for the range of planar random walks, 2009

928 **Ulrich Bunke,** Index theory, eta forms, and Deligne cohomology, 2009

927 **N. Chernov and D. Dolgopyat,** Brownian Brownian motion-I, 2009

926 **Riccardo Benedetti and Francesco Bonsante,** Canonical wick rotations in 3-dimensional gravity, 2009

925 **Sergey Zelik and Alexander Mielke,** Multi-pulse evolution and space-time chaos in dissipative systems, 2009

924 **Pierre-Emmanuel Caprace,** "Abstract" homomorphisms of split Kac-Moody groups, 2009

923 **Michael Jöllenbeck and Volkmar Welker,** Minimal resolutions via algebraic discrete Morse theory, 2009

922 **Ph. Barbe and W. P. McCormick,** Asymptotic expansions for infinite weighted convolutions of heavy tail distributions and applications, 2009

921 **Thomas Lehmkuhl,** Compactification of the Drinfeld modular surfaces, 2009

920 **Georgia Benkart, Thomas Gregory, and Alexander Premet,** The recognition theorem for graded Lie algebras in prime characteristic, 2009

919 **Roelof W. Bruggeman and Roberto J. Miatello,** Sum formula for SL_2 over a totally real number field, 2009

918 **Jonathan Brundan and Alexander Kleshchev,** Representations of shifted Yangians and finite W-algebras, 2008

917 **Salah-Eldin A. Mohammed, Tusheng Zhang, and Huaizhong Zhao,** The stable manifold theorem for semilinear stochastic evolution equations and stochastic partial differential equations, 2008

916 **Yoshikata Kida,** The mapping class group from the viewpoint of measure equivalence theory, 2008

915 **Sergiu Aizicovici, Nikolaos S. Papageorgiou, and Vasile Staicu,** Degree theory for operators of monotone type and nonlinear elliptic equations with inequality constraints, 2008

914 **E. Shargorodsky and J. F. Toland,** Bernoulli free-boundary problems, 2008

913 **Ethan Akin, Joseph Auslander, and Eli Glasner,** The topological dynamics of Ellis actions, 2008

912 **Igor Chueshov and Irena Lasiecka,** Long-time behavior of second order evolution equations with nonlinear damping, 2008

911 **John Locker,** Eigenvalues and completeness for regular and simply irregular two-point differential operators, 2008

910 **Joel Friedman,** A proof of Alon's second eigenvalue conjecture and related problems, 2008

TITLES IN THIS SERIES

909 **Cameron McA. Gordon and Ying-Qing Wu,** Toroidal Dehn fillings on hyperbolic 3-manifolds, 2008

908 **J.-L. Waldspurger,** L'endoscopie tordue n'est pas si tordue, 2008

907 **Yuanhua Wang and Fei Xu,** Spinor genera in characteristic 2, 2008

906 **Raphaël S. Ponge,** Heisenberg calculus and spectral theory of hypoelliptic operators on Heisenberg manifolds, 2008

905 **Dominic Verity,** Complicial sets characterising the simplicial nerves of strict ω-categories, 2008

904 **William M. Goldman and Eugene Z. Xia,** Rank one Higgs bundles and representations of fundamental groups of Riemann surfaces, 2008

903 **Gail Letzter,** Invariant differential operators for quantum symmetric spaces, 2008

902 **Bertrand Toën and Gabriele Vezzosi,** Homotopical algebraic geometry II: Geometric stacks and applications, 2008

901 **Ron Donagi and Tony Pantev (with an appendix by Dmitry Arinkin),** Torus fibrations, gerbes, and duality, 2008

900 **Wolfgang Bertram,** Differential geometry, Lie groups and symmetric spaces over general base fields and rings, 2008

899 **Piotr Hajłasz, Tadeusz Iwaniec, Jan Malý, and Jani Onninen,** Weakly differentiable mappings between manifolds, 2008

898 **John Rognes,** Galois extensions of structured ring spectra/Stably dualizable groups, 2008

897 **Michael I. Ganzburg,** Limit theorems of polynomial approximation with exponential weights, 2008

896 **Michael Kapovich, Bernhard Leeb, and John J. Millson,** The generalized triangle inequalities in symmetric spaces and buildings with applications to algebra, 2008

895 **Steffen Roch,** Finite sections of band-dominated operators, 2008

894 **Martin Dindoš,** Hardy spaces and potential theory on C^1 domains in Riemannian manifolds, 2008

893 **Tadeusz Iwaniec and Gaven Martin,** The Beltrami Equation, 2008

892 **Jim Agler, John Harland, and Benjamin J. Raphael,** Classical function theory, operator dilation theory, and machine computation on multiply-connected domains, 2008

891 **John H. Hubbard and Peter Papadopol,** Newton's method applied to two quadratic equations in $\mathbb{C}^2$ viewed as a global dynamical system, 2008

890 **Steven Dale Cutkosky,** Toroidalization of dominant morphisms of 3-folds, 2007

889 **Michael Sever,** Distribution solutions of nonlinear systems of conservation laws, 2007

888 **Roger Chalkley,** Basic global relative invariants for nonlinear differential equations, 2007

887 **Charlotte Wahl,** Noncommutative Maslov index and eta-forms, 2007

886 **Robert M. Guralnick and John Shareshian,** Symmetric and alternating groups as monodromy groups of Riemann surfaces I: Generic covers and covers with many branch points, 2007

885 **Jae Choon Cha,** The structure of the rational concordance group of knots, 2007

884 **Dan Haran, Moshe Jarden, and Florian Pop,** Projective group structures as absolute Galois structures with block approximation, 2007

883 **Apostolos Beligiannis and Idun Reiten,** Homological and homotopical aspects of torsion theories, 2007